Some aspects of the botany of the Shetland Islands

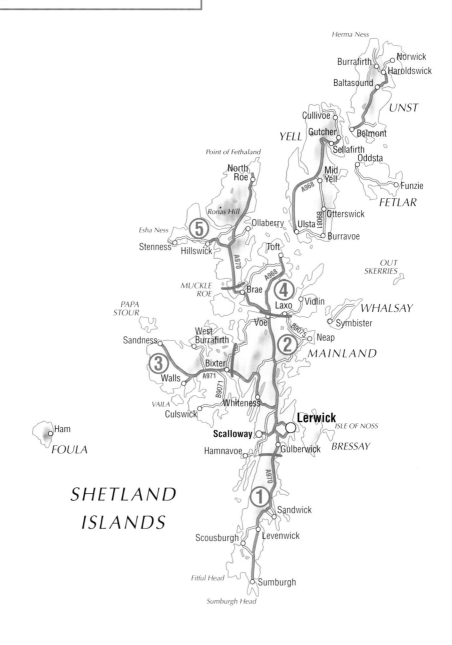

Key to the major divisions of Mainland

(1) South Mainland (2) Central Mainland

(3) West Mainland (4) North Mainland

(5) Northmavine

Herma Ness

Norwick
Burrafirth
Haroldswick
Baltasound

UNST

Cullivoe
Gutcher
Belmont
Sellafirth
Oddsta

YELL

Point of Fethaland

North
Roe

Mid
Yell

Funzie

FETLAR

Ronas Hill

Ollaberry
Otterswick
Ulsta
Burravoe

Esha Ness

Stenness
Hillswick

Toft

OUT
SKERRIES

A970

A968

MUCKLE
ROE

Brae

(4)

Laxo

Vidlin

WHALSAY

PAPA
STOUR

Voe

B9075

Neap

Symbister

West
Burrafirth

(2)

MAINLAND

Sandness

Bixter

(3)

Walls

A971

B9071

Whiteness

VAILA

Culswick

Lerwick

ISLE OF NOSS

Scalloway

BRESSAY

Ham

Hamnavoe

Gulberwick

FOULA

(1)

Sandwick

SHETLAND
ISLANDS

Scousburgh

Levenwick

A970

Fitful Head

Sumburgh

Sumburgh Head

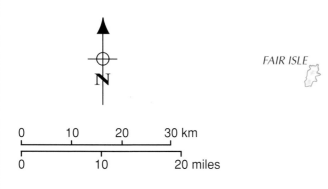

N

| 0 | 10 | 20 | 30 km |

| 0 | 10 | 20 miles |

FAIR ISLE

Some aspects of the botany of the Shetland Islands

Walter Scott

W. Scott

Look at

Perforate

St John's

Wort

Published by the au
Scalloway, Shetland I
2011

Published by Walter Scott, 'New Easterhoull', Castle
Street, Scalloway, Shetland, ZE1 0TP

First published: Scalloway 2011

ISBN 978-0-9567836-0-8

Relevant titles co-produced by the author

A Check-list of the Flowering Plants and Ferns of the Shetland Islands.
Scalloway and Oxford 1969. (With R. C. Palmer.)
The flowering plants and ferns of the Shetland Islands.
Lerwick 1987. (With R. [C.] Palmer.)
Rare plants of Shetland. [Lerwick] 2002. (With Harvey, P.,
Riddington, R., and Fisher, M.)

Camera-ready copy (in Times New Roman) and scan imagery
(including the front cover image) by the author

Front cover image: Mainland Hawkweed (*Hieracium
subtruncatum*). See p.66 and Plate 8

Printed by The Orcadian, Hell's Half Acre,
Hatston, Kirkwall, Orkney, KW15 1DW

Contents

Preface

This book, as its name suggests, covers some aspects of the flora of Shetland. Four of these features are included, and, although the pagination is continuous, each has its own title and can be regarded as a separate document in its own right.

The opening part is a list of species which have been reliably recorded since records began nearly 250 years ago. Each species is accompanied by a notes on its status, habitats, and frequency, etc. Many of the notes are quite short, but othere are more detailed, especially where rare, scarce, or threatened plants are involved. The information in these longer notes is in many cases appearing in print for the first time.

The second part is a dichotomous key to the twenty-seven species of *Hieracium* (hawkweeds) known to grow in Shetland, including two which are no longer believed to occur in the wild, *H. maritimum* and *H. hethlandiae.* This genus is of considerable significance within our flora; of the eighteen species which are endemic to our area, sixteen belong to section Alpestria.

Scanned images of herbarium sheets of very typical material of all of our twenty-two endemic plants of the Asteraceae form the third part of this book, and are spread over three genera as follows: *Taraxacum* (3), *Pilosella* (1), and *Hieracium* (18).

The fourth part is a comprehensive bibliography of the Shetland flora.

The writer wishes to thank The Orcadian, Kirkwall, Orkney, for printing the book, and to Ashworth Maps and Interpretation Limited, Glasgow, for the map showing the divisions of Mainland.

Walter Scott
'New Easterhoull'
Castle Street
Scalloway
Shetland
ZE1 0TP

February 2011

Dedicated to all of those botanists—from James Robertson in 1769 to
Richard Palmer in the latter half of the twentieth century—
and to the many others who have added so much to our
knowledge of the flora of the Shetland Islands

Part 1

A new checklist of the flowering plants, ferns, and fern allies of the Shetland Islands

The first comprehensive records of Shetland plants, as defined by the above title, are two documents produced by James Robertson who, in 1769, visited Orkney and Shetland in the summer of that year. His first Shetland document lists 252 names. This figure rises to 297 by the addition of forty-five species to his second document, a tabular flora of Shetland and other northern and western islands of Scotland which he had also surveyed about the same period. For reasons unknown to the writer fifteen names in the first list do not appear in his second document. Many of his records are the earliest notices for Shetland of species which, prior to the recent discovery of Robertson's manuscripts in the Signet Library in Edinburgh, had been assumed to have been first recorded by Thomas Edmondston, and others. Indeed, only *Calluna vulgaris, Narthecium ossifragum,* and *Cochlearia officinalis* in Robertson's lists had been certainly recorded by others before his visit. Unfortunately, his documents contain scientific names only—no localities or notes on frequency or habitats—and we are left to assume that, in the case of rarities, Robertson had probably seen them in the places which were to become very familiar to later recorders. His more interesting (and confirmed) records include *Apium inundatum, Arctium nemorosum* (as '*Arctium lappa'*), *Geranium robertianum, Odontites vernus, Osmunda regalis, Ranunculus bulbosus,* and *Zostera marina.* Although Robertson visited Unst he seems not to have examined the Keen of Hamar and thereby overlooked our two great rarities, *Arenaria norvegica* subsp. *norvegica* and *Cerastium nigrescens* var. *nigrescens.* The pleasure of discovering these fell to the very young Edmondston nearly seventy years later. Nevertheless, Robertson's manuscripts are very important early botanical documents which give us a clear picture of our flora nearly two-and-a-half centuries ago. (Henderson & Dickson (1994) provide us with an interesting account of his Scottish travels, along with a reconstruction of his tabular flora.)

In 1806 the youthful Yorkshire naturalist Charles Fothergill spent a good part of that year in studying the natural history of Orkney and Shetland. His catalogue (Fothergill 1806), which combines in one document the plants seen by him in the two areas, is the earliest, most comprehensive and annotated botanical document covering both groups of islands. Unlike Robertson's Shetland documents, Fothergill's catalogue supplies both Latin and English names, with many notes on habitats, distribution, localities, and interesting miscellaneous observations on a number of species. The total number of names in Fothergill's Orkney and Shetland document is 420, a commendable figure even after allowing for a number of wrong or improbable records. Of this total, only seventy-two are specifically listed, or are unequivocally implied, as occurring in Shetland. Included in these are eleven first records for Shetland, and about thirteen errors or probable errors. Fothergill, obviously, would have seen numerous additional plants in Shetland; unfortunately, his combining of the two counties in a single document has led to many general statements in which the species concerned is not expressly recorded or implied as occurring in one county or the other. A separate list from each county would have been so much more useful, especially at this early date.

In 1969, exactly 200 years after Robertson's list was written, the first comprehensive and annnotated checklist of the Shetland flora appeared. *A Check-list of the Flowering Plants and Ferns of the Shetland Islands*, compiled and published by Richard Palmer and Walter Scott, covered 681 taxa and provided scientific and English names, and a brief note on each entry. Unlike Robertson, who had to break new ground, the compilers were able to draw on a number of publications, especially those by T. Edmondston (1845), Tate (1866), Craig-Christie (1870), Beeby (1887–1909), West (1912), Druce (1922–1925), Spence (1979), and on the compilers' own work. The period covered by these investigations extended from the late 1830s to the late 1960s. No checklists covering the whole of Shetland were published during this time. (Robertson's list was unknown to the compilers when their checklist was produced.) The 1969 list was updated twice, Scott & Palmer (1980, 1999), the latter also served as an update to Scott & Palmer (1987), the current standard work on Shetland plants. Details of these earlier workers' chief publications are given at the end of this list. The present checklist is being compiled without the help of Richard Palmer whose death in 2005 marked the end of a lifetime in the study of the local flora, not only in the field but in libraries and herbaria. His thirty-three visits to Shetland began in 1955 and ended in 1999 when failing health made it less than easy for him to undertake the oft-repeated journey from Oxford. A major proportion of his Shetland material is in the South London Botanical Institute. There is also a good representation in the Fielding-Druce Herbarium in the University of Oxford.

This checklist covers 899 taxa divided between two main headings: native (451 taxa) and non-native (448 taxa), the latter (collectively referred to as aliens) divided between established aliens and casuals. The ten hybrids between natives and aliens, and between aliens, are regarded as natives as they are likely to have arisen *de novo* in the area. It is interesting to note that the total recorded flora is almost equally divided between natives and aliens.

Natives (388, plus 10 species believed to be extinct in the wild. (The total of 398 taxa includes 22 endemics)	398
Native hybrids: between natives (37); native hybrids where one parent is not on record (6); between natives and aliens (8); and between aliens (2)	53
Established aliens (including 55 Taraxaca)	161
Casuals (94 of generally unknown origin, and 193 of generally agricultural, horticultural (including casuals of amenity reseeding), or of bird-seed origin	287
Total	899

In this checklist natives are unmarked, endemics are indicated by an obelus (†), and all established aliens and casuals are signalled by an asterisk (*). For statistical purposes a taxon is accorded its higher current status where more than one applies. Ten native taxa now believed to be extinct appear in square brackets. These are *Atriplex laciniata, Crambe maritima, Eryngium maritimum, Glyceria declinata, Hieracium hethlandiae, H. maritimum, Minuartia rubella, Polygonum oxyspermum, Pseudorchis albida,* and *Pyrola* sp.

The following are the now generally accepted definitions of the two primary status categories, native and alien. A **native** species 'is one which arrived in the study area without intervention by man, whether intentional or unintentional, having come from an area in which it is native *or* one which has arisen *de novo* in the study area'. *Calluna vulgaris* has been here for a very long time and is a good example of a native species. Equally native was the single plant of *Crambe maritima* on a Cunningsburgh beach *c.*1991 (for one year only), and which must have arrived naturally as a seaborne seed from a native site outwith Shetland or from an undiscovered native site within the county. The seas and tides around us must carry many seeds of coastal plants around our own shores and from farther afield. An **alien** species 'is one which was brought to the study area by man, intentionally or unintentionally, even if native to the source area *or* one which has come into the area without man's intervention, but from an area in which it is present as an introduction'. Within a study area a plant may be native in one place, but alien in another. These two definitions (of native and alien) are from Preston, Pearman & Dines (2002) and provide a workable division between these two main categories.

The writer has divided aliens into established aliens and casuals. The chief features of an **established alien** are its non-native status, its long persistence in a given site if undisturbed, its often close proximity to human habitation or activity (crofting, farming, gardening, etc.), its propensity to spread into natural vegetation, and, in the case of (mostly annual) arable or garden weeds, its ability to colonise newly bared soil from long-buried seeds. Most of these are well-established throughout the county, and only a few are listed as weakly established. Some are widespread and popular garden plants such as *Crocosmia × crocosmiiflora, Hyacinthoides × massartiana,* and *Narcissus* sp. Others include once-common annual weeds of arable ground, notably *Galeopsis tetrahit* sensu lato*, Lamium purpureum, Matricaria discoidea, Polygonum boreale, Raphanus raphanistrum,* and *Spergula arvensis.* Changing agricultural practices have now largely replaced the widespread patchwork of small fields of oats, potatoes, etc, with grass-based systems. From 1971 to 2008 (both years included) the area under barley, oats, turnips, swedes, kale, cabbage, and potatoes fell from 1,320 to 148 hectares. During the same period the area under grass (excluding rough grazings) rose from 6,998 to 23,920 hectares (Shetland Islands Council (2009)). There are sixty-three recorded Taraxaca; five of these are native, three probably native, while the remainder are treated as established aliens. See p. 60 for further information.

A *casual* is also non-native but differs chiefly in being unable to perpetuate itself either vegetatively or by seed. Included here are the (often) annual species of disturbed ground, rubbish-tips, neglected areas around harbours, building sites, gardens, etc. They rarely survive more than a year either because seed may not ripen or germinate, or because the habitat becomes overgrown. The waifs and strays of our flora, they are generally of unknown provenance. A good example is the single plant of *Parentucellia viscosa* which turned up in 1998 at Aith, West Mainland. Also regarded as casuals are taxa which have clearly originated from agricultural operations (relics of sown hayfields, etc.), as are contaminants in mixtures for reseeding roadside embankments and other areas where new grass cover is required. Other casuals are of garden origin, namely clump-forming perennials, trees and bushes, etc., which have been planted in the wild or appear as garden throw-outs or escapes. Casuals must all have one feature in common, the inability to regenerate either by vegetative means or by seed, whether an annual herb or a long-lived tree.

The foregoing information on categories and definitions is based on Macpherson, *et al.* (1996), Preston *et al.* (2002), and Stace, *et al.* (2003). Sequence, nomenclature, and English names conform very closely to Stace (1997) except *Taraxacum* and *Hieracium* where Sell & Murrell (2006) are followed, and *Trichophorum* where Jermy, *et al.* (2007) are observed. Names of authorities are not given. New records of interest and notes on other species appearing for the first time in a local botanical publication include the date of discovery and the finder's name, the latter given directly or by implication, or by his initials. All place-names appear on the current Ordnance Survey 1:25,000 maps or have their positions indicated by reference to a named feature on these maps. Localities are usually listed under the five major areas of Mainland (the main island) namely, South, Central, West, and North Mainland, and the parish of Northmavine, the most northerly portion of North Mainland. Northmavine is is a large area surrounded by the sea except for a tiny neck of land (Mavis Grind) at its S end where its connection to the remainder of North Mainland is little more than wide enough to carry the main road. Because of its very interesting flora and diverse geology, along with its almost complete separation from the rest of Mainland, Northmavine is treated as a recording division in its own right. Yell, Unst, Fetlar, and all the smaller islands are referred to individually. (See map facing the title-page and showing the major recording divisions of Mainland.)

Huperzia selago subsp. **selago**. *Fir Clubmoss*. Frequent and widespread on heaths and moors and associated bare peaty or stony places; rarely in noticeable quantity except on the upper slopes of Ronas Hill, Northmavine, and its neighbouring summits.
Lycopodium clavatum. *Stag's-horn Clubmoss*. Fine heathy ground. Rare. Despite seven new sites in recent years from Central, West, and North Mainland, and the first records from Whalsay (2004, H. Thomson) and Fetlar (2006, A. Williamson), its long-term decline continues. In some old sites it has either vanished or become very scarce. Happily, it is not extinct in the Hill of Easterhoull/Bersa Hill area, Central Mainland as claimed by Scott & Palmer (1987); in 2004 it was refound by W. Moore.

Diphasiastrum alpinum. *Alpine Clubmoss*. An arctic-alpine plant of fine heathy ground. Local, except on Ronas Hill, Northmavine, and the plateau to the N where it is relatively frequent. Also on Sandness Hill, West Mainland; on several hills W of Dales Voe, North Mainland; on Hurda Field, near Mavis Grind, Northmavine (1987, WS); and in Muckle Roe (2002, WS). There are old and unconfirmed records from Fetlar and Unst.

Selaginella selaginoides. *Lesser Clubmoss*. A widespread arctic-alpine plant, often common in damp heathy-mossy places, and on the drier hummocky parts of marshes. It is particularly abundant on the serpentinite grass-heath to the N of Baltasound, Unst.

Isoetes lacustris. *Quillwort*. An occasional to frequent plant of stony or gravelly loch bottoms, sometimes with the following species in the same loch.

I. echinospora. *Spring Quillwort*. Less often encountered than the preceding and with a preference for the West and North Mainland. It favours silty-bottomed lochs.

Equisetum fluviatile. *Water Horsetail*. A fairly frequent and often abundant plant of watery places. Loch margins (where it often forms large stands), sluggish streams, ditches, and swampy areas.

E. × litorale (*E. fluviatile* × *E. arvense*). *Shore Horsetail*. An occasional to locally frequent plant of burnsides, roadsides, etc. This hybrid was discovered in a number of widely scattered places by R. C. Palmer, and there can be little doubt that it has been widely overlooked. It is surprising that W. H. Beeby, the first person to detect this hybrid in Britain, seems not to have recognised it in Shetland.

E. arvense. *Field Horsetail*. A frequent plant of dry roadsides, undercliffs, waste and weedy places, etc., as well as being a persistent weed in cultivated ground, especially in sandy areas. The brown erect fertile stems look very different from the rather straggly, somewhat bushy barren stems which develop later in the season.

E. sylvaticum. *Wood Horsetail*. An occasional and graceful horsetail of moist or wet places: streambanks, marshes, bogs, undercliffs, etc. Scattered throughout Shetland.

E. palustre. *Marsh Horsetail*. Common in a variety of watery places such as marshes, burnsides, ditches, and the like. Once seen on the dry stone walls of a ruined crofthouse at Catwalls, Tingwall, Central Mainland.

Ophioglossum azoricum. *Small Adder's-tongue*. A small and easily overlooked fern of fine coastal pasture. Often recorded from small offshore islands where surveys may tend to be more thorough than in the less well-defined parts of our extensive coastline. Very rarely seen away from the coast, as near the large standing stone near Lund, Unst (1995, BSBI field meeting; 2009, L. Farrell).

Botrychium lunaria. *Moonwort*. Formerly a frequently encountered fern of short dry, often coastal or rocky, grassland, especially on serpentinite or limestone soils; also on turfy roadside verges. Although widely distributed, it is declining and, as elsewhere in the country, rarely seen in significant numbers in any one place.

Osmunda regalis. *Royal Fern*. Long known from holms in five lochs in West Mainland. Tiny plants and sporelings are also found on these holms, as well as by the shore of one of the lochs (South Loch of Hostigates). In 1991 L. A. Inkster and WS found a single plant twelve cm high on a holm in the Loch of Voxterby (still there in 2003), and in 2003 WS found another plant twenty cm high on a tiny, reef-like, wave-swept 'holm' in Sulma Water. In 1978 a small plant was found by the stream from Culeryin to The Vadills but was not seen again. Unfortunately, for one reason or another, these occurrences outside of the main sites (from where the spores must have originated) do not result in new and successful colonies, though the sporelings must play an important part in maintaining the plant in its established locations. This splendid fern once grew, long ago, near Sandwick, Unst, but was eventually exterminated by man. An even earlier record is provided by J. Robertson in 1769, but, unfortunately, he gave no locality.

Hymenophyllum wilsonii. *Wilson's Filmy-fern.* A frequent fern of damp mossy hollows among rocks or under boulders, moist streambanks, and on steep, upland, heathy pastures with a northerly or easterly aspect. It is often in the company of mosses and is usually stunted and browned, rarely showing its dark green and translucent fronds to perfection except in deep shade. This easily overlooked and probably much under-recorded fern still grows by the Burn of Skaw, Unst, from where it was recorded around 1840 by T. Edmondston.

Polypodium vulgare. *Polypody.* Frequent in rocky places and ravines, and on heathery burnsides and sea-banks, etc. Also commonly seen on the walls of long-abandoned crofthouses, as at Culsetter, near Mavis Grind, North Mainland, and at 'Heogravalta', near Everland, Fetlar. Polypody is very tolerant of exposure and is often seen in a very dwarfed condition; in more sheltered places, however, it can be quite luxuriant.

P. × mantoniae (*P. vulgare × P. interjectum*). Only once recorded, 'on rocks, Mavis Grind [North Mainland], large and small plants together, the former in sheltered crevices' (1953, A. H. G. Alston). This record was never confirmed and the rocks where it grew may have been blasted away a long time ago. The second parent has not been noted in Shetland and could be awaiting discovery.

Pteridium aquilinum. *Bracken.* Frequent in the northern parts of West Mainland, and by the N side of Ronas Voe, Northmavine (where there are several large patches). Well-drained peaty grassland or heathery slopes, or among scree, and by the sides of the more sheltered *voes*. Recently, an unrecorded site very far from its nearest neighbour came to light: a small patch on a steep rocky bluff by the W side of Hamna Dale, S of Lunning, North Mainland (2006, WS). Extinct in Foula.

Phegopteris connectilis. *Beech Fern.* Scattered stations in sheltered rocky places in crevices or under hollows, usually in small quantity, in the Ronas Hill area N to the Beorgs of Uyea, Northmavine. A single station is known in West Mainland: by Robie Glen's Loch, E of Sulma Water. In 2003 L. A. Inkster found one plant, probably casual, on the rock-armour embankment on the seaward side of the North Atlantic Fisheries College (now the NAFC Marine Centre), Scalloway. The following year it had been unwittingly eradicated. The origin of this plant is unknown, as are the origins of other ferns which have appeared here, notably *Phyllitis scolopendrium* and *Polystichum aculeatum*. The site seems most unsuitable for ferns in general, being drenched in sea spray in strong winds.

Oreopteris limbosperma. *Lemon-scented Fern.* A local fern of Central and North Mainland, very rare elsewhere. Dry rocky or bouldery places, ditches, roadside verges and slopes, etc., often as single tufts. Recently detected in Mousa (2002, R. Norde).

Phyllitis scolopendrium. *Hart's-tongue.* Very rare. One small plant by a loch in West Mainland. In 2005 a single plant was seen on the walls of Fort Charlotte, Lerwick, by L. A. Inkster, followed by another, elsewhere on the walls, by LAI and

WS. Also in 2005 another plant, probably casual, was found by LAI under a drain cover in a path on the seaward side of the North Atlantic Fisheries College (now the NAFC Marine Centre), Scalloway, just behind the rock-armour embankment. Still there (2008, WS). Records from Cunningsburgh, South Mainland, and the Burn of Sundibanks, near Scalloway, both before1845, have never been reliably confirmed.

Asplenium adiantum-nigrum. *Black Spleenwort*. Essentially a plant of rock-crevices, sometimes in scree. Scattered throughout Shetland, especially in the rockier parts of West and North Mainland, and on the Unst and Fetlar serpentinite. Very rare in the peaty island of Yell.

A. marinum. *Sea Spleenwort*. A truly maritime, evergreen fern of spray-washed sea-cliffs, rock-fissures and cool moist hollows around many parts of the coastline; rarely a little way inland (as on the Snarra Ness peninsula, West Mainland), and usually in small numbers. A fine station, however, exists at Esha Ness, Northmavine, at the Dale of Ure, many hundreds of plants on shaded cliffs and in hollows near a natural arch (1999, WS).

A. trichomanes subsp. **quadrivalens**. *Maidenhair Spleenwort*. Shady crevices of limestone or sandstone rocks. Local in parts of Central Mainland and northern West Mainland, very rare in South Mainland (Fladdabister) and in Unst (Loch of Cliff). Recently found near the Burn of Fitch, E of Scalloway (2003, P. V. Harvey) and on the walls of Fort Charlotte, Lerwick (2005, L. A. Inkster and WS).

A. viride. *Green Spleenwort*. Very rare. Discovered by C. W. Peach in 1864 on Muckle Heog, Unst. Still there, under boulders and in rock-crevices, but sparingly.

A. ruta-muraria. *Wall-rue*. Rare. Very dry rocks (where it occurs sparingly and is easily overlooked) and on walls. By the Loch of Clousta, the North Loch of Hostigates, and to the NE of the latter. Also near planted shrubs by the roadside near Vementry (Mainland), and on the Snarra Ness peninsula. All of these sites are in West Mainland. Also in Vementry island. In Scalloway it is found on walls about 'Gibblestone House', and in Whalsay it is frequent on the old walls of the Midden Court behind 'Symbister House' (now a school) and on a roadside wall to the north.

Athyrium filix-femina. *Lady-fern*. A frequent fern of sheltered burnsides, ravines, etc., and among rocks where it is often severely stunted. Withering early in autumn. The plumose form which once grew at Boddam is still in cultivation in Scalloway.

Gymnocarpium dryopteris. *Oak Fern*. Local in North and Central Mainland, very rare or rare elsewhere. Chiefly a fern of rock-crevices or among stones, boulders or scree, rarely on streambanks. It also likes old stone walls or other man-made stone structures (as on the old fish-drying area at The Ness, Burra Firth, Unst). Rarely in noticeable quantity. Several new sites have been found recently by L. A. Inkster.

Cystopteris fragilis. *Brittle Bladder-fern*. A rare, small and delicate fern of rocky places by burns on limestone in South and Central Mainland, on sandstone on high ground in Foula, and on man-made stone structures such as the broch at Clickimin, Lerwick (noted in 1980 and still there in 1991). Very rare in West Mainland. It has recently been found at Fladdabister, South Mainland (2003, L. A. Inkster), and in surprising quantity on a wall at Fort Charlotte, Lerwick (2005, LAI and WS).

***Polystichum aculeatum**. *Hard Shield-fern*. Probably casual. A single plant on the rock-armour embankment at the North Atlantic Fisheries College (now the NAFC Marine Centre), Scalloway (1998, L. A. Inkster and WS). It was still there in the spring of 1999 but had gone by the autumn. A second plant appeared, in 2000 (LAI), close to the site of the first. It survived for several years before it, too, disappeared.

Dryopteris filix-mas. *Male-fern*. Among rocks, by burns and on sheltered sea-banks, by roadsides and in old quarries, etc. Generally distributed in Shetland apart from some of the smaller islands. Often singly, but frequent in Crossie Geo, N of Sandwick, South Mainland (a crisped congested form), and on the S side of Firths Voe, near Mossbank, North Mainland.

D. **affinis** group. *Scaly Male-fern*. Rare. Rocky or gravelly roadside embankments, low sea-banks, among boulders, etc., in a handful of places from Channerwick, South Mainland, to Sobul, Unst. Fraser-Jenkins (2007), in a very detailed paper on this difficult group, recognises *D. affinis* subsp. *affinis, D. cambrensis*, and (doubtfully) *D. borreri* from Shetland. The writer prefers, for the time being, a single heading pending more research locally.

D. **dilatata**. *Broad Buckler-fern*. This is one of our most widespread and common ferns. Burnsides, crags, scree, holms in lochs, heathy and moory places (luxuriant in moist moorland hollows), etc. Also on man-made stone structures, particularly on long-abandoned crofthouses and boundary walls.

D. **expansa**. *Northern Buckler-fern*. A rare fern, primarily among boulders and rocks in the Ronas Hill area, Northmavine: above the Black Butten at *c.*350 m (1991, L. A. Inkster and WS); on a flattish summit *c.*600 m SE of the Hill of Burriesness (1994, WS); on the N side of The Roonies, NNE of Swabie Water (1994, LAI and WS); and *c.*400 m NNW of Abram's Ward (2000, LAI and WS). There are only two other records. In 1976 a single tuft was found by a stream in one of the woods at Kergord Plantations, Weisdale, Central Mainland. (There is doubt as to whether it grew in the Leegarth or the Lindsay Lee wood.) This was the first record for Shetland, and the fern may well have been native. It has not been observed at Kergord again. In 1980 a small quantity was found in a peaty hollow W of Vatster, Central Mainland, but in 2008 WS noted that it was in very poor condition as a result of grazing and seemed to be on the verge of extinction.

Blechnum spicant. *Hard-fern*. A common fern of damp heathery or grassy-heathery places such as burnsides, heathy or rocky slopes, and on coastal banks.

*Picea sitchensis. *Sitka Spruce*. Casual. Very rarely in wild or semi-wild places.

*Larix decidua. *European Larch*. Casual. Very rarely in wild or semi-wild places.

*L. kaempferi. *Japanese Larch*. Casual. Very rarely in wild or semi-wild places.

Juniperus communis subsp. **nana**. *Common Juniper*. An arctic-alpine shrub. Dry stony heaths, crags, burnsides, sea-banks, and holms in lochs. Rare or local (often only a few plants at each place) and seldom in significant quantity except in Fair Isle, and by the Burn of Swabiewater, North Roe, Northmavine (1994, S. C. Jay); not uncommon in Vementry and Muckle Roe. Several interesting new sites have recently come to light: rocky sea-cliffs on the W side of Papa Little (1991, R. C. Palmer); on a rock pinnacle near the Boo Stacks, W side of Burra Firth, Unst (1992, WS); many patches on the sides of a *geo* at the N end of Longa Berg, Skelda Ness, West Mainland (1993, L. Farrell and WS); and in a *geo* at HU496053, W of the entrance to Gloup Voe, Yell (2000, WS). Juniper is probably continuing to decline in Shetland, and is almost certainly extinct in Foula. Some early writers on Shetland appear to have found it relatively easily, and this leaves little doubt that it was once more widespread.

J. communis, form intermediate between subsp. **communis** and subsp. **nana**. Very rare and certainly gone from three of its four recent sites. Single bushes have been found by or near roadsides in three places in South Mainland: above the SW side of the Loch of Spiggie (1967, H. J. M. Bowen); near the A970/B9122 junction above Channerwick (*c.*1963, J. R. Colvin); and above Upperton, Levenwick (1989, JRC). The first two were exterminated by roadworks, the third by a landslip. In 1992 WS found a small bush on moorland on Fitsyi Field, Fetlar, but has been unable to find it again. There is little doubt that this is the same taxon which T. Edmondston, around 1840, recorded from Cunningsburgh and Dunrossness, both in South Mainland. The writer has a fine bush (raised at Cambridge from a Spiggie cutting) in his garden.

Nymphaea alba subsp. **occidentalis**. *White Water-lily*. This fine plant occurs as a native in six lochs or swampy pools in the West Mainland, and has been introduced in several suitable places within and outwith its native area. In all cases the transplanted material came from West Mainland. First noted, in 1774, by G. Low.

Caltha palustris. *Marsh-marigold*. Common and often abundant, brightening many a burn or loch margin in early summer.

***Trollius europeaus**. *Globeflower*. Casual. Merely a rare garden outcast.

***Aconitum napellus**. *Monk's-hood*. Casual. An outcast or garden relic.

***Anemone nemorosa**. *Wood Anemone*. A well-established alien in Peggy's wood, Kergord Plantations, Weisdale, Central Mainland, where several large patches exist.

***A. coronaria**, cultivar. *Poppy Anemone*. Casual. Garden outcast.

Ranunculus acris. *Meadow Buttercup*. Abundant and widespread in a large variety of habitats from watery ditches, burnsides, loch margins, lowland meadows and pastures, to dry rocky grassland and fine coastal turf.

***R. repens**. *Creeping Buttercup*. A common and well-established alien of damp cultivated ground in the crofting and farming districts, on waste ground and beaches, and around long-abandoned crofthouses; also a persistent weed in gardens.

R. bulbosus. *Bulbous Buttercup*. Rare. Dry pastures (often sandy) by or near the coast. It can be found at St Ninian's Isle, South Mainland; Bousta, Sandness, West Mainland; West Sand Wick, Yell; and at Skeo Taing, Unst. Recently, it has turned up in cliff-top pasture NNE of the Loch of Norby, Sandness, West Mainland (1997, WS) and, in Papa Stour, abundantly in the Biggings area (1997, WS), at Kirk Sand (2002, WS), and near a sandy beach at the W side of West Voe (2002, WS).

R. flammula subsp. **flammula**. *Lesser Spearwort*. A common and very variable plant. Found in a variety of watery places: ditches, bogs and marshes, sides of lochs, etc. Also in bare, damp, peaty or stony places by the coast as an extreme form approaching but not quite matching subsp. *minimus* which was recorded from Shetland in 1953 but has apparently not been seen since.

***R. aconitifolius**. *Aconite-leaved Buttercup*. Casual. A garden outcast or deliberate introduction.

R. ficaria subsp. **ficaria**. *Lesser Celandine*. A popular and frequent plant of damp grassland, burnsides (often with yellow irises), sea-banks and rocky hollows, about churches and cemeteries, and in gardens. A welcome splash of colour in spring.

R. hederaceus. *Ivy-leaved Crowfoot*. In streams and ditches and on muddy or sandy loch margins from near Lerwick to Fair Isle. Local to frequent in South Mainland, sometimes abundant and choking sluggish watercourses.

9

R. baudotii. *Brackish Water-crowfoot*. Very local. A plant of lochs and pools in the extreme S of South Mainland: Scat Ness; Lochs of Spiggie, Hillwell, and Gards, etc. It flourishes in Maa Loch, Vementry, and is recorded from the Loch of Grunasound, West Burra.

R. trichophyllus. *Thread-leaved Water-crowfoot*. Very local and, like the preceding, mainly in the extreme S of South Mainland: Lochs of Spiggie, Huesbreck, and Gards, and in pools among the dunes behind the Bay of Quendale, etc. Also in the Lochs of Kirkigarth and Bardister, Walls, and recently found in small quantity in the Loch of Grunnavoe, near Walls (2003, WS), all in West Mainland. An outlying site occurs in Haaf Gruney, Unst.

****Aquilegia** sp. *Columbine*. Casual. Plants close to *A. vulgaris*, including possible hybrids or cultivars of this or other species, are sometimes seen as garden escapes or outcasts.

Thalictrum alpinum. *Alpine Meadow-rue*. An arctic-alpine, generally frequent but unrecorded from much of South Mainland. Wet stony or rocky places, such as flushes, loch margins, damp burnsides, etc., especially on limestone or serpentinite.

****Papaver pseudoorientale**. *Oriental Poppy*. Casual. A very showy garden outcast or straggler; probably also sometimes a deliberate introduction.

****P. somniferum** subsp. **somniferum**. *Opium Poppy*. Casual. Two plants among oats, Mail, Cunningsburgh, South Mainland, 1963; and as a field weed near Crawton, Sandness, West Mainland (1995, BSBI field meeting). Elsewhere it has been recorded as an obvious garden escape or outcast.

****P. rhoeas**. *Common Poppy*. Casual. Nowadays a very rare casual of recently reseeded roadside verges and embankments: a few plants at the head of the East Voe of Scalloway (1991, M. C. Johnson); one plant at the Blett junction, Cunningsburgh, South Mainland (1992, L. A. Inkster and WS). Long ago it certainly seems to have been a cornfield weed. C. Fothergill, during his tour in 1806, records it from cornfields and claims that it was commoner here than in Orkney. Edmondston (1845) lists it thus, 'Fields, Skaa, Unst; Northmavin', the habitat presumably applying to both stations.

****P. dubium** subsp. **dubium**. *Long-headed Poppy*. Once a well-established alien in arable fields on sandy ground in southernmost South Mainland and (similarly) in the northern parts of Yell and Unst. Grass-based systems have largely replaced the traditional practices and the poppy is now rare in these areas. Two instances of long-buried seeds being brought to the surface and germinating have been recorded in recent years: large numbers of plants on newly-made verges of the upgraded A970 between Boddam and the Pool of Virkie, South Mainland (1987, WS); hundreds of plants on the site of a new house, 'Hjemdaal', Upper Scalloway (1996, Z. Gilfillan).

****Meconopsis cambrica**. *Welsh Poppy*. A well-established alien. This very popular garden plant is often seen near gardens as an escape or outcast; also in old quarries and other places where garden rubbish has been dumped.

****Glaucium flavum**. *Yellow Horned-poppy*. More likely to be a casual than a native from seaborne seed. There is only one record, from 'Sulam Voe, Northmavin' by T. Edmondston (1845). This would indicate that it grew by the west side of Sullom Voe.

****Fumaria muralis**. *Common Ramping-fumitory*. Casual. Once found, in 1968, in a potato-patch, Cutts, Trondra.

*F. officinalis subsp. **officinalis**. *Common Fumitory*. A well-established alien. Dry, sandy arable ground in the S half of South Mainland (sometimes common) and in similar places elsewhere as at Bruntland, Bressay, and in the sandy NE corner of Yell. Otherwise it is rare or unrecorded, particularly in much of West and North Mainland, and Northmavine.

*Ulmus glabra. *Wych Elm*. Casual. Very rarely seen in a semi-wild situation.

*Humulus lupulus. *Hop*. Casual. Merely a rare garden straggler. It has strayed through a stone-built garden wall at 'Kelda', Baltasound, Unst, since 1962 or possibly much earlier.

*Urtica dioica. *Common Nettle*. A well-established alien. Commonly seen about houses and crofts (especially crofthouse ruins from where it is rarely absent), waste grassy or rubbly places, higher parts of beaches, etc.

*U. urens. *Small Nettle*. A well-established alien. A small annual nettle which is local or frequent throughout Shetland as an arable weed on light sandy soils (most notably in southernmost South Mainland), on sandy seashores and in waste places, etc.

*Soleirolia soleirolii. *Mind-your-own-business*. Casual. On the remains of the sunken garden, 'Leagarth House', Houbie, Fetlar. First noticed in 1967, it is likely to have been present much earlier; in 2000 it was still surviving in very small quantity, but in 2008 WS thought it had vanished under lush vegetation.

Betula pubescens subsp. **tortuosa**. *Downy Birch*. One of our rarest native shrubs and found in only five sites in the Ronas Hill and North Roe areas of Northmavine: on two holms in a loch at the Clubbi Shuns; the holm in the larger of the Inniscord Lochs; on a steep and almost inaccessible rock-face at The Trip; and on a sea-bank at the Slocka beach. In 1928 it was found on a holm in one of the lochs at the Many Crooks, N of Ronas Hill. The birch was still surviving in 1953 but by 1959 it had died, probably through grazing pressure.

Corylus avellana. *Hazel*. An even rarer shrub. It is known from only two sites: a rocky ravine at Catfirth, South Nesting, Central Mainland, and on a holm in Punds Water, Northmavine.

*Chenopodium vulvaria. *Stinking Goosefoot*. A bird-seed casual in the garden of 9 Knab Road, Lerwick (1997, I. Clark).

*C. album. *Fat-hen*. A well-etablished alien. Local in sandy arable ground, mainly in South Mainland. Recent records include: Levenwick (1996, R. C. Palmer), and below Gord Farm, near Hillwell (1997, WS), both in South Mainland; Bruntland, Bressay (1998, WS); and Leogh, Fair Isle (1998, N. Riddiford). Elsewhere it is seen rarely as a garden weed or on waste disturbed ground.

Atriplex prostrata sensu stricto. *Spear-leaved Orache*. Very rare on or near seashores in Fair Isle, the only place where it is known as a native. Two sites were found by R. C. Palmer in 1977: near Shieldi Geo but seen for a year or two only, and near the Skadan lighthouse where it still survives. N. Riddiford, in 1999, discovered a third site: near the pier at the South Harbour. A casual plant closely resembling this was once found, in Lerwick, at Holmsgarth Road (1991, WS).

A. × gustafssoniana (*A. prostrata × A. longipes*). *Kattegat Orache*. Recorded from four sites, from Aith Voe, Cunningsburgh, South Mainland, to Mid Yell Voe.

A. glabriuscula. *Babington's Orache*. Apparently local. Recorded from a handful of stations from Fair Isle to Balta Sound, Unst, including the Muckle Ossa, off Esha

Ness, Northmavine. Much over-recorded in the past but perhaps the opposite applies now.

A. glabriuscula × **A. praecox**. Once recorded: on the shore near Sellafirth, Yell.

A. × taschereaui (*A. glabriuscula* × *A. longipes*). *Taschereau's Orache*. Material thought to be close to this has been collected at Channer Wick, South Mainland.

A. praecox. *Early Orache*. This small, early flowering species is frequent on damp gravelly shores of sheltered *voes* and associated semi-enclosed *houbs* and *vadills*; often among seaweed in the upper part of the intertidal zone.

A. littoralis. *Grass-leaved Orache*. Very rare. Known only from the shore and adjoining rough herbage at Boddam, South Mainland, where, in 1985, it was noticed by D. H. Dalby. It is probably a recent seaborne arrival as there is no earlier reference to it in this well-worked area. In 2003 P. V. Harvey counted well over 100 plants. The writer, in 2009, commented, 'There is more of this now at Boddam, on both sides of the *voe* and at its head, scattered plants and groups of many individuals.'

*****A. patula**. *Common Orache*. A well-established alien. Frequent as a weed of arable ground, especially on sandy soils in the S of South Mainland. Occasional elsewhere in crops, about houses and gardens, waste places, etc. It likes well-manured ground.

[**A. laciniata**. *Frosted Orache*. Believed to be extinct. A few plants were found by W. H. Beeby in 1899 on a sandbank at the shore near Clavel, South Mainland. The plant did not appear there the following year.]

Salicornia europaea sensu stricto. *Common Glasswort*. Very rare. A plant of quiet muddy seashores, sometimes in the intertidal zone. Recorded by Edmondston (1845) in an aggregate sense from Dales Voe, near Lerwick; Sullom Voe, North Mainland; and Baltasound, Unst. It has never been confirmed in the first two places, leaving Balta Sound (both sides of the inlet) as the only station in Shetland. The Unst plant (assuming there is only one species there) is placed under this heading on the basis of fresh material, collected in November 1976, and confirmed by T. G. Tutin.

Suaeda maritima. *Annual Sea-blite*. Very rare. Quiet, muddy seashores. Nowadays seen only on both sides of Balta Sound, Unst. Recorded from Dales Voe, near Lerwick (pre-1845), and Haroldswick, Unst (1920 or 1921), but no later reports exist from either place. More recently, in 1976 and 1977, found at The Houb, Swinister, North Mainland.

*****Amaranthus** sp. A bird-seed casual. Very sparingly on disturbed ground near new roadworks, 'South Taing', Cott, Weisdale, Central Mainland (2009, R. Leask). This is almost certainly **A. retroflexus** (common amaranth), according to E. J. Clement.

*****Claytonia perfoliata**. *Springbeauty*. A very rare garden casual, found at Mid Walls, West Mainland (*c*.1981), and at Scalloway (1986, 1987).

*****C. sibirica**. *Pink Purslane*. A well-established alien in several of the plantations at Kergord, Weisdale, Central Mainland. A popular garden plant, it is seen elsewhere as a casual straggler or outcast on damp barish ground in the inhabited areas.

Montia fontana. *Blinks*. A common plant which occurs in a variety of wet, rocky or stony, places, on peat, and in flushes, damp grassland, cultivated fields and gardens. Four subspecies, all based on seed-coat features, are recognised from Britain. Two of these are recorded from Shetland. Subsp. **fontana** is the common and widespread form. Subsp. **variabilis** has one record: the slopes of Ruska Lee, W of North Haven, Fair Isle (1957, A. Currie). It has not been seen there again.

Arenaria serpyllifolia subsp. **serpyllifolia**. *Thyme-leaved Sandwort.* It is too soon to regard this as a presumably extinct native, even though it was last seen, in 1989, in the Scalloway station This small annual plant favoured two distinct habitats and was in small numbers in most, if not all, of its stations. The main area was the sandy arable ground, associated pastures, sand-dunes, and dry roadsides in the extreme S end of South Mainland, at Ringasta, Hillwell, Quendale, Exnaboe, and Sumburgh. It was last seen in this habitat, in 1979, at Sumburgh. The other habitat was dry limestone outcrops: near Scalloway Junior High School; Brough, and by the Loch of Benston, both in South Nesting, Central Mainland, and by Papil Water, Fetlar. In 1989 R. C. Palmer found a single plant in sown grass by the Unst Leisure Centre, evidently casual.

A. **norvegica** subsp. **norvegica**. *Arctic Sandwort.* This is one of our national rarities. It is confined to sparsely vegetated serpentinite fellfield in three areas in Unst only: Sobul; from E of Crussa Field to the Keen of Hamar, including from SW of to SE of Muckle Heog; and on the Hill of Clibberswick. In the third site it seems to be very scarce; in fact, there is only one recent record from here, a single plant (1989, R. C. Palmer).

Honckenya peploides. *Sea Sandwort.* A frequent plant of sandy-shingly seashores, sometimes extending to adjacent dunes or rocky sand-filled niches. Very rarely on skerries, as on Muckle Skerry, near Out Skerries (1968, J. Blance).

[**Minuartia rubella**. *Mountain Sandwort.* Believed to be extinct. It once grew to the N of Baltasound, Unst, almost certainly on serpentinite fellfield. Undoubted plants were collected in 1840, and again in 1886 (when some of the material was localised to the Wick of Hagdale), but since then there have been no specimens or photographs to back up the few later reports from both N and S of Baltasound.]

Stellaria media. *Common Chickweed.* A very common plant of cultivated ground (including gardens where it can be a troublesome weed), on bare waste ground, also by the coast on shingle beaches, and luxuriant on offshore holms and stacks.

*****S**. **holostea**. *Greater Stitchwort.* A well-established alien in Kergord Plantations, Weisdale, Central Mainland, since the mid 1950s and likely to have been present much earlier. Still locally plentiful in Peggy's, Leegarth, and Lindsay Lee woods.

*****S**. **graminea**. *Lesser Stitchwort.* A well-established alien in one place: Boddam, South Mainland, in wet pasture bordering a (now) bypassed section of the A970, first noticed in 1957 and still there in 1999 when the patch occupied fourteen square metres. Otherwise a scarce casual by roadsides (sometimes persisting for a time), in reseeded pastures, and (rarely) in cultivated fields.

S. **uliginosa**. *Bog Stitchwort.* A common plant of low-lying watery places: ditches, burns, pools, and flushes.

*****Cerastium tomentosum**. *Snow-in-summer.* Casual. A popular, white-flowered, mat-forming, garden plant. It is sometimes seen as a straggler or outcast, or as a deliberate introduction among rocks near houses where it can be quite persistent.

C. **nigrescens** var. **nigrescens**. *Shetland Mouse-ear.* This is the most celebrated plant in the county and the finest of T. Edmondston's discoveries. It is found in Unst only, on dry, sparsely vegetated serpentinite fellfield in the Keen of Hamar area, and in the same habitat, but in dry or wet places, from SW of to SE of Muckle Heog. Shetland mouse-ear once grew between Uyeasound and Baltasound, in 1894, but has not been seen since. It was probably found in the Sobul area where its companion rarity elsewhere in Unst, *Arenaria norvegica* subsp. *norvegica*, still grows. A narrow-leaved form occurs on Muckle Heog and (very rarely) on the Keen of Hamar. Brysting (2008), in a very detailed and rather technical paper on the *C. alpinum* complex, to which the Shetland plant belongs, explains that the Unst taxon cannot be upheld as an endemic species to Shetland despite such distinctive features as its abundant glandular pubescence and purple colouration. She says that 'Shetland plants have not diverged enough to be separated as a distinct species', and concludes by stating that 'the Scottish *C. nigrescens* shares its evolutionary history with *C. nigrescens* from Shetland and mainland Norway, and this non-arctic taxon is clearly separated from the arctic taxon, *C. arcticum*, with which it has previously been considered conspecific'. (Brysting, therefore, does not regard *C. arcticum* (to which British (including Shetland) plants have often been referred) as occurring in Britain.) Her paper should be read in conjunction with Brummitt, *et al.* (1987). Stace (2010) places the Unst plant under the endemic var. *nigrescens* of *C. nigrescens*. Whether or not this marks the end of the debates over the taxonomy and nomenclature of the Unst plant, which have appeared fairly regularly in the literature ever since Edmondston brought his plant to public notice in the late 1830s, remains to be seen.

C. fontanum. *Common Mouse-ear.* Common in a wide variety of habitats, both wet or dry: pastures; marshes; rocky places; roadsides; cultivated ground and bare rubbly soil; and on the walls of ruined crofthouses, etc. Very variable. The Shetland plant appears to be subsp. **vulgare**, but more study of the species should be undertaken over the whole range of its habitats.

*****C. glomeratum**. *Sticky Mouse-ear.* A well-established alien. Frequent in crofting and farming areas, on bare rubbly ground, etc. It likes a rich soil and is not averse to being trampled.

C. diffusum. *Sea Mouse-ear.* Common around the coast in rock-crevices, on dry, often sandy, grassy banks, and among sand-dunes where the relative shelter allows it to flourish.

Sagina nodosa. *Knotted Pearlwort.* Occasional and rarely in quantity. It prefers limestone and other basic soils. Marshy places, flushes, wet mossy turf (among sand-dunes, etc.), dry serpentinite debris, loch margins, and (rarely) in saltmarshy turf. In 1806 C. Fothergill found it to be 'pretty frequent' in the 'bogs of Shetland'.

S. subulata. *Heath Pearlwort.* Frequent in dry rocky or stony places: crags, stony heathy pastures, gravelly roadside verges, etc. It is largely represented by the glabrous form in Shetland.

S. procumbens. *Procumbent Pearlwort.* A common plant of dry or wet rocky or stony places (streamsides, loch margins, shingle beaches, flushes), by ditches and in dry or wet turf, and a garden weed on damp soil and in flower pots and paving cracks.

S. maritima. *Sea Pearlwort.* Frequent on coastal rocks and in adjacent short turf and on barish cliff-top soil; also on coastal stonework. Rarely found some way from the coast.

*****Spergula arvensis**. *Corn Spurrey.* A well-established alien. Formerly common in nearly every district in cultivated ground (oats, turnips, etc.) and about crofts and farms, especially on poor soil; also appearing on excavated soil originating from areas of former cultivation. Now much scarcer owing to changing agricultural practices.

Spergularia media. *Greater Sea-spurrey.* A local to frequent plant of saltmarshes and brackish turf, rarely on muddy beaches. It prefers wetter places than those favoured by the following species, the two rarely growing together.

S. marina. *Lesser Sea-spurrey.* Local, but not uncommon by the E side of Shetland. Bare, often rocky, small offshore holms and spray-swept skerries; barish cliff-top pastures; saltmarshy turf (often by pools); shell-sandy pastures; coastal stonework; and rarely on muddy shingle, or by inland roadsides up to 1.6 km from the sea.

Lychnis flos-cuculi. *Ragged-Robin.* Widespread in damp meadows and pastures, and very fine on some grassy holms (Calf of Little Linga, Whalsay, 2003, WS), etc.

*Agrostemma githago. *Corncockle.* Status uncertain; probably best regarded as a former casual of cultivated fields. There have been no records since around 1840; Edmondston (1845) speaks of it as an imported field weed. The only known specimen is an undated one by him from Unst in the herbarium of the University of Manchester. Tate (1866) records it from 'cultivated lands at Tingwall', N of Scalloway, but he seems not to have seen it himself.

Silene uniflora. *Sea Campion.* A frequent coastal species of shingly beaches, sea-cliffs, and rocky offshore islands; also seen some way inland as on barren serpentinite debris in Unst and Fetlar, and (very rarely) in craggy places. Raeburn (1891), a famous mountaineer of his time, recorded this, as 'bladder campion', on the top of the Lyra Skerry, off the W side of Papa Stour.

S. acaulis. *Moss Campion.* An arctic-alpine plant which descends to sea-level in Shetland. It prefers limestone and granite and is locally frequent by or near the coast on bare gravelly or rocky slopes, and on outcropping rocks in fine, short pasture. Rarer inland in stony or rocky places, but a notable component of bare serpentinite debris in Unst.

*S. latifolia. *White Campion.* An extremely rare casual of (usually) sandy arable ground in the S of South Mainland, and on waste ground elsewhere, most often as single specimens. Recorded from Hillwell (1965, 1969, 1997); near Exnaboe (1966, 1969); on the Lerwick rubbish-dump (by the Loch of Clickimin, 1955); and on a heap of soil, Quendale Lane, Lerwick (1976). The following records have been made recently: a clump on the gravelly verge of the A970 near the Pool of Virkie, South Mainland (1996, R. C. Palmer); one plant by the edge of a field of *Avena strigosa* by the road below 'Busta', Fair Isle (2004, N. Riddiford).

S. × hampeana (*S. latifolia × dioica*). Twice recorded, in the 1960s, from near Exnaboe, South Mainland, and Lerwick.

S. dioica. *Red Campion.* A frequent and characteristic species of the coastline. Typically found on sea-bird cliffs, grassy sea-banks, and the upper parts of shingly beaches, commonly dark-flowered, handsome and luxuriant (the so-called subsp. *zetlandica*). Farther inland it is seen about houses, by old walls, and in disused quarries.

*S. gallica. *Small-flowered Catchfly.* A casual with a single record: in sown grass by the roadside at the head of the East Voe of Scalloway, 1991, a single plant.

*Persicaria campanulata. *Lesser Knotweed.* Casual. Garden outcast or deliberate introduction.

*P. bistorta. *Common Bistort.* A local and weakly established alien of garden origin. Often by burnsides (sometimes among irises), in ditches, or near houses.

P. vivipara. *Alpine Bistort.* An occasional, but sometimes locally frequent arctic-alpine. Dry coastal pastures (stony or heathy), especially on the west side of Shetland; also on rocky or stony high ground near and on the summits of Fair Isle, Foula, and Ronas Hill, Northmavine.

P. amphibia. *Amphibous Bistort.* Frequent in South Mainland, West Mainland, Unst, the N part of Yell, and in Fetlar, rare elsewhere. The aquatic form occurs in lochs, marshy or swampy places, wet meadows, on dry roadsides, and as an arable weed.

*P. maculosa. *Redshank.* A well-established alien, and a local to frequent species of arable ground, rarely by loch margins or on beaches in or close to farming areas. It occasionally appears on heaps of soil which have been excavated from areas where crops had once grown, and on levelled soil of similar origin, as in 2009 (WS) when it appeared in great abundance during the construction of the Endavoe housing scheme, East Voe, Scalloway.

*P. lapathifolia. *Pale Persicaria.* A bird-seed casual. Garden of 9 Knab Road, Lerwick (1998, I. Clark). Fair Isle Bird Observatory (1999, N. Riddiford).

*Fagopyrum esculentum. *Buckwheat.* Another bird-seed casual. Garden of 9 Knab Road, Lerwick (1998, I. Clark). One plant on gravelly ground at the West Burrafirth ferry jetty, West Mainland (2003, G. Laurenson).

[**Polygonum oxyspermum**. *Ray's Knotgrass.* Believed to be extinct. It is correctly recorded on only one occasion: sands of Burra Firth, Unst, where, in 1868, it was collected by A. Craig-Christie. Earlier, Edmondston (1845) had written: 'seashores, common' [in Shetland generally] but he may have included other members of the *Polygonum aviculare* group.]

*P. arenastrum. *Equal-leaved Knotgrass.* A well-established alien which is found frequently around crofts and farms (on tracks, about sheep-pens and field gates, and on trampled ground generally), and as an arable weed of (mainly) sandy soils.

*P. aviculare sensu stricto. *Knotgrass.* A well-established alien. Very rare. Known only in and near arable ground in Fair Isle. Once recorded as a bird-seed casual: weed in garden of 9 Knab Road, Lerwick (1998, I. Clark).

*P. boreale. *Northern Knotgrass.* A well-established alien. Formerly common in arable fields (especially cornfields), and around crofts and farms generally. Now much less common owing to grass-based systems having largely replaced the small weedy fields of oats, potatoes, turnips, etc., which were once common in every district throughout Shetland.

*Fallopia japonica. *Japanese Knotweed.* A well-established alien. A fairly frequent garden escape or outcast which has become established in many places near houses, notably by burns and in neglected areas where it can spread unhindered. The distinctive var. *compacta* occurs in three places in Yell: near the cemetery, Reafirth, Mid Yell, and on the sea-bank below; a large colony in a field behind a house by the B9083, Cullivoe, close to a ruined house, 'Ark', on the other side of the road, but nearly all enclosed (1997, R. C. Palmer). Still at all three sites in 2008 (WS) and showing signs of spreading at the Cullivoe station.

*F. sachalinensis. *Giant Knotweed.* A well-established alien. Known only from Houbie, Fetlar, where it has been around since the early part of the twentieth century. Only a small amount now remains (following attempts to keep it in check), and this should be saved from extinction before it is too late, if only for its possible link with the nearby 'Leagarth House' garden.

***F. baldschuanica**. *Russian-vine.* Casual. Garden straggler on a wall or two at 'Leagarth House', Houbie, Fetlar.

***F. convolvulus**. *Black-bindweed.* A local casual of arable ground (often sandy); on rubbish-dumps; and rarely by weedy roadsides and in gardens, etc. Nearly always in small numbers but in 1997 R. C. Palmer found 'a fair number of plants' in sandy arable ground by the Loch of Hillwell, South Mainland. It had been seen here in 1965 and 1982 and is perhaps, at this place, a rather weakly established alien.

***Rheum × hybridum**. *Rhubarb.* Casual. Widely grown in Shetland and often seen as an outcast or relic of cultivation in or near the inhabited areas.

Rumex acetosella subsp. **acetosella**. *Sheep's Sorrel.* A common plant of the poorer soils: bare peat, poor arable ground, rocky places, moors, and on gravel.

R. acetosa subsp. **acetosa**. *Common Sorrel.* Common in damp grassy places (field-borders, sides of ditches and burns), holms in lochs, and on sea-cliffs, etc.

R. acetosa subsp. **hibernicus**. Frequent on serpentinite debris N of Baltasound, Unst. It has also been found in a similar habitat near the Loch of Winyadepla, Fetlar, where confirmation of its continued existence would be desirable.

R. longifolius. *Northern Dock.* A common dock in the inhabited areas, also on sea-cliffs, small offshore islands, holms in lochs, seashores, etc.

R. × propinquus (*R. longifolius × R. crispus*). Probably widely overlooked. Only a handful of very old records exist, and only one modern record: garden of 'Headlands', Exnaboe, South Mainland (2003, P. V. Harvey, A. Lockton, and S. Whild). A serious study of our docks has never been undertaken and is now badly needed.

R. × hybridus (*R. longifolius × R. obtusifolius*). Probably widely overlooked. Like the preceding, there are very few records and all are old or very old.

R. crispus subsp. **crispus**. *Curled Dock.* Frequent in cultivated ground and waste places. By the coast (on seashores, among rocks, and on sea-cliffs) it is largely, if not always, replaced by the equally frequent subsp. **littoreus**.

*****R. obtusifolius**. *Broad-leaved Dock.* A frequent and, if alien, a well-established plant of waste or neglected places in the inhabited and cultivated areas. It was considered by R. C. Palmer that this was more likely to be an alien than a native in Shetland, bearing in mind its liking for areas of past and present human activity.

Oxyria digyna. *Mountain Sorrel.* A rare arctic-alpine confined to the Ronas Hill, Northmavine, area where it favours wet (sometimes scree-filled) coastal gullies, cliffs, and low sea-banks; rarely by rocky streams a little way inland. It can be found here and there, usually in small quantity, from the burn between Birka Water and Lang Clodie Loch, to sea-cliffs near the waterfall from the last-named loch, and S to just N of The Priest, on the N side of Ronas Voe. It also grows in one or two sites by the S side of Ronas Voe.

Armeria maritima subsp. **maritima**. *Thrift.* A very abundant coastal species of cliff-top pastures, among rocks, on holms and skerries, in saltmarshes and brackish turf, and on man-made structures by the shore. Also inland on the Unst serpentinite debris, the granite debris of the Ronas Hill summit, Northmavine, and occasionally on gravelly roadside verges. Robert Louis Stevenson, in 1869, noted its presence on the walls of Fort Charlotte, Lerwick.

Elatine hexandra. *Six-stamened Waterwort.* Very rare. Recorded from two pairs of nutrient-rich lochs: the Lochs of Spiggie (2004) and Brow (1981), South Mainland, and from the Lochs of Kirkigarth (2002) and Bardister (1997), West Mainland. The dates are those of the latest sightings. It grows on a muddy-sandy bottom and seems always to be submerged in Shetland.

*Hypericum perforatum. *Perforate St John's-wort.* Casual. There are four records, all on reseeded embankments or verges by the A970 and all in South Mainland: about sixty plants on the verge of the (old) A970 at the bridge over the Burn of Claver, Channerwick (2002, J. Halcrow), very few in 2004, none in 2005, a few in 2006, and fewer than a score in 2009 (all J. Nicolson); several patches (totalling *c.*200 plants) NW of Hoswick at HU405248 (2002, JH), very few in 2004, none in 2005, a few in 2006, and *c.*1,400 stems (2009, JN); several patches just S of the bridge over the Burn of Channerwick where only two plants were seen *c.*2003 (2007, T. Russell), no later count; and one or two plants NW of the Loch of Fladdabister (2007, R. Norde). The huge increase noted at the second station in 2009 is remarkable.

*H. tetrapterum. *Square-stalked St John's-wort.* Casual. Garden of 'Brakes', S of Boddam, South Mainland (2003, I. Bairnson).

H. pulchrum. *Slender St John's-wort.* This is a local to frequent plant of dry rocky-heathery burnsides and sea-banks, rocky or stony heaths, and of the serpentinite grass-heath of Unst.

*Malva nicaeensis. *French Mallow.* Casual, once recorded. A single large plant among potatoes in the garden of 44 Castlepark, Lerwick (1999, T. Slater).

*Sidalcea sp. *Greek Mallow.* Casual. Popular garden plants rarely seen outside enclosures. A clump was found in 1965 by a stream at West Sandwick, Yell, and persisted until at least 1986 when it was in poor condition. It has not been recorded from there since. The Yell plant has been tentatively referred to **S. hendersonii**.

Drosera rotundifolia. *Round-leaved Sundew.* Frequent and widespread, but only occasionally locally common. Boggy sphagnous places on moors and heaths. The writer is aware of only two occurrences of fully opened flowers: between Septa Field and the Twart Burn, W of Heylor, Northmavine (1 August 1995, WS); about the N end of Bays Water, W of Busta, Delting, North Mainland (6 July 2006, WS).

D. anglica. *Great Sundew.* An apparently very rare plant. Boggy sphagnous moors; preferring wetter places than those frequented by the preceding species. Known with certainty from two sites in Northmavine: S side of Roer Water (last seen in 1984); by the W side of the Burn of Roerwater just above and below where it is joined by the Burn of the Twa-roes where, in 2003, a total of fifty-three clumps were seen by P. V. Harvey, A. Lockton, and S. Whild. Very old records from Fetlar and Yell remain unconfirmed.

Viola riviniana. *Common Dog-violet.* A common species of dry grassy banks both coastal and inland, in pastures (often heathy), by rocky burnsides, on crags, etc.

V. canina subsp. **canina**. *Heath Dog-violet.* Frequent on the Unst and Fetlar serpentinite and local to occasional on dry coastal banks and slopes elsewhere; also among the sand-dunes of the Links of Quendale, South Mainland (1996, R. C. Palmer). Once thought to be almost confined to Unst, *V. canina* has recently been found in a number of coastal sites (including the islands of Papa Stour, Vementry, and Muckle Roe). It will almost certainly turn up elsewhere in Shetland.

V. palustris subsp. **palustris**. *Marsh Violet.* Common in damp, rough, rushy or heathy pastures and in boggy or marshy ground.

*****V. cornuta**. *Horned Pansy.* Formerly a garden outcast by and below the road below the Asta farmhouse, N of Scalloway. It was first noticed in 1955 and was still present in 1976, but by 1983 it had vanished.

*****V. tricolor** subsp. **tricolor**. *Wild Pansy.* Formerly a well- established alien in arable ground (particularly in the sandy areas of South Mainland) and in fallow fields; among sand-dunes (perhaps marking former cultivation); in waste places; and as a garden weed. Nowadays it is much scarcer owing to the reduction in widespread small-scale arable farming. Our sand-dune plant does not appear to be subsp. *curtisii* which is recorded from as near as the N end of Orkney.

V. tricolor × arvensis (*V. × contempta*). Very rare. Only two records exist, both from South Mainland; hayfield between Scatness and Sumburgh Airport, 1966, and in a cornfield in the Souther House area, above the Loch of Spiggie two years later.

*****V. arvensis**. *Field Pansy.* Nowadays, this is likely to be a rather weakly established alien. It was formerly a well-established alien in arable ground in three extensive sandy areas: the S end of South Mainland; the NE corner of Yell; and Norwick, Unst. The loss of habitat in these areas has severely reduced its numbers. The latest sightings are as follows: a few plants near 'Betty Mouatt's Cottage', Scatness, South Mainland (1999, R. Riddington); near the 'Haa of Houlland', Yell (1961, WS); and in four fields (common in one), Norwick (1990, WS). It is also seen rarely as a casual.

Populus tremula. *Aspen.* A scarce native of rocky sea-banks and crags. Found in six sites: Bay of Brenwell, near Sandness, West Mainland; Ness of Houll, opposite Muckle Roe, North Mainland; Ness of Isleburgh, near Mavis Grind; near Slocka, and E of the Grud Burn, both on the N side of Ronas Voe—these three sites are all in Northmavine; and by the E side of Whale Firth, Yell.

*****Salix pentandra**. *Bay Willow.* Casual. Very rarely planted outside enclosures, as by the road NE of the Strand Loch, Tingwall, Central Mainland (1994, R. C. Palmer).

*****S. fragilis**. *Crack-willow.* Casual. Very rarely planted outside enclosures. A bush in poor condition grew, in 1985, at the mouth of the burn at Swining, near Vidlin, North Mainland.

*****S. × mollissima** (*S. triandra × S. viminalis*). *Sharp-stipuled Willow.* Casual. In 1978 a willow believed to be this hybrid grew in Burns Lane, Lerwick.

*****S. purpurea**. *Purple Willow.* Casual. Very rarely planted outside enclosures.

*****S. daphnoides**. *European Violet-willow.* Casual. Formerly planted in two places: by a roadside, Veensgarth, Tingwall, Central Mainland; and by the Ham Burn, Foula.

*****S. viminalis**. *Osier.* Casual. Rarely planted outside enclosures. Three 'bowed and twisted trees' by wall, Lunna, N of Vidlin, North Mainland (1996, R. C. Palmer).

*****S. × sericans** (*S. viminalis × S. caprea*). *Broad-leaved Osier.* Casual. A deliberate introduction, as at the lower end of the Burn of Quoys, South Nesting, Central Mainland, where it has grown for over a century; also elsewhere in Shetland.

*****S. × calodendron** (*S. viminalis × S. caprea × S. cinerea*). *Holme Willow.* Casual. Planted by a burn at North Dale, Fetlar, where it has been known for a long time.

*****S. × stipularis** (*S. viminalis × S. caprea × S. aurita*). *Eared Osier.* Casual. Planted by a stream at Geosetter, S of Boddam, and similarly at Wester Quarff, both in South Mainland; and at the Pund of Grutin, Dales Voe, Delting, North Mainland.

*S. × **smithiana** (*S. viminalis* × *S. cinerea*). *Silky-leaved Osier*. Casual. A very popular willow in Shetland and one which often survives by an abandoned crofthouse long after the roof has gone.

*S. **caprea**. *Goat Willow*. Casual. This is very rarely planted outside enclosures.

S. **cinerea** subsp. **cinerea**. *Grey Willow*. One of our rarest native willows. Known only from the two fenced-off, contiguous holms in Mousavord Loch, West Mainland.

S. **cinerea** subsp. **oleifolia**. *Grey Willow*. Rare. Confined to sea-banks or sea-cliffs in Fair Isle; at Quey Firth, Northmavine; and near Aywick, Yell. Also on holms in eight lochs, one in Vementry island, the other seven on the North Roe plateau N of Ronas Hill, Northmavine. The best site by far in Shetland is the large holm in Sandy Water, North Roe. (The records in Scott & Palmer (1987) from Houlma Water and the Neeans, both near West Burrafirth, West Mainland, are errors for *S. × laurina*.)

S. × **multinervis** (*S. cinerea* × *aurita*). A rare willow hybrid which occurs on the holms in Roer Water, Tonga Water, and the larger of the Inniscord Lochs. Also by the shore of the last-named loch, and by the burn running into the Brettoo Loch, all in North Roe. All of its stations are in Northmavine.

*S. × **laurina** (*S. cinerea* × *S. phylicifolia*). *Laurel-leaved Willow*. Casual. Often seen outside enclosures in West Mainland. It is well suited to a watery habitat.

S. **aurita**. *Eared Willow*. Our most frequent large native willow, especially in West and North Mainland, local or occasional elsewhere. Mainly by burns and on holms in lochs, but also in damp pastures, among rocks, etc.

S. × **ambigua** (*S. aurita* × *S. repens*). A local hybrid in West and North Mainland, very rare elsewhere. Heathery or rocky streamsides, damp pastures, and holms in lochs. A good site for it is the tiny holm close inshore at the N end of Sulma Water, West Mainland (2003, WS).

S. **repens**. *Creeping Willow*. The most frequent and widespread native willow in Shetland. Heathy (occasionally sandy) pastures both damp or dry, steep sea-banks, crags (sometimes forming large patches), and holms in lochs. It is particularly abundant around the margin of Little Holm, in the Loch of Watlee, Unst (2003, WS).

S. **lapponum**. *Downy Willow*. Very rare. Known only from a holm in Moosa Water, North Roe, Northmavine. Downy willow is very much an arctic-alpine shrub of the Scottish mountains; the Shetland site, at *c*.140 m, may be the lowest in the country. The habitat, too, may be unique in Britain. The writer has a fine plant, raised from a Moosa Water cutting, in his garden in Scalloway.

S. **herbacea**. *Dwarf Willow*. Not uncommon on stony summit plateaux (fellfield) and exposed rocks on hills over *c*.180 m, especially on the dry granite debris of Ronas Hill, Northmavine, and its neighbouring summits. An arctic-alpine, it is much scarcer at lower levels, but is frequent on the North Roe plateau, and descends almost to sea-level at Fethaland, both in the same division. It is also found in Fair Isle, Foula, Yell, Fetlar, Unst, and in a number of places in Mainland.

*Sisymbrium orientale**. *Eastern Rocket*. Casual. In 1956 this grew on the Lerwick rubbish-dump (by the Loch of Clickimin) and persisted for at least five years or until the early 1960s when the site was turned over to pasture. In 1963 a patch was found on the replacement tip (on the site of the North Loch) but was not recorded again.

*S. **officinale**. *Hedge Mustard*. A casual found in two places in Lerwick in 1969: on the rubbish-dump (on the site of the North Loch), and on waste ground, Chromate

20

Lane. There are three new records: in pavement cracks by house 'Annslea', Scalloway (1996, WS); roadside verge above Channerwick, South Mainland (1997, L. A. Inkster); and a garden weed at Lower Stonybreck, Fair Isle (N. Riddiford, 2001).

*__Descurainia sophia__. _Flixweed._ A casual with just one record: a single plant in 1982 in a sandy arable field by the Loch of Hillwell, South Mainland.

*__Alliaria petiolata__. _Garlic Mustard._ Another casual with one record. In 1971 one plant was found as a garden weed in Twageos Road, Lerwick.

*__Arabidopsis thaliana__. _Thale Cress._ A very rare casual weed, mainly of gardens, and first noted in 1928 at Brough Lodge, Fetlar. It was seen at Lerwick in 1958, and more recently in and near Scalloway (as by a former fish-processing factory near Blythoit, East Voe, 2009, WS).

*__Erysimum cheiranthoides__. _Treacle-mustard._ A very rare casual. Discovered in 1990 as a garden weed at 'Busta', Fair Isle.

*__Hesperis matronalis__. _Dames's-violet._ Casual. A very sweetly scented and popular garden plant, usually called sweet rocket. It often occurs as an outcast on waste ground, rubbish-tips, rubbly embankments, and weedy sea-banks below villages; also an escape in the vicinity of gardens.

*__Malcolmia maritima__. _Virginia Stock._ Merely a rare casual of garden origin with a few old records from rubbish-dumps and waste ground.

*__Matthiola longipetala__. _Night-scented Stock._ Casual. Once recorded from Lerwick.

*__Barbarea vulgaris__. _Winter-cress._ A rare casual with only three recent records: nearly twenty plants on a grassy roadside by the Vandlope turn, Scousburgh, South Mainland (1996, R. C. Palmer); one plant in a ditch on the S side of the road, Roesound, Muckle Roe (2004, WS), still there two years later; one plant in grass within the new Whiteness cemetery, Central Mainland (2008, L. A. Inkster and WS). In 1974 R. C. Palmer found a single plant near the post office, Scousburgh. The two records from here may be unrelated; if not, it is a remarkable case of persistence for a plant which is supposedly a casual.

*__B. verna__. _American Winter-cress._ Apparently a genuine casual, not deliberately grown. One record (in (1962): garden border, Hoswick, Sandwick, South Mainland.

*__Rorippa nasturtium-aquaticum__ sensu stricto. _Water-cress._ A well-established alien. This is essentially a plant of fairly nutrient-rich ditches, streams and lochs in the sandy far south of South Mainland, occurring here and there from the Lochs of Spiggie and Brow (sparingly by the former) to the Loch of Hillwell and the Links of Quendale, sometimes locally abundant. It also grows at the Loch of Gards, Scatness (2006, BSBI field meeting), and about Sumburgh. A few miles to the N water-cress grows in roadside ditches S of Sand Lodge, Sandwick; at Sandwick hamlet; and possibly between the two. It was at Sand Lodge that A. Craig-Christie in 1868 recorded it for the first time for Shetland. The first West Mainland record was made recently: a large patch in a watercourse just W of the old schoolhouse on the N side of Scutta Voe, N of Gruting (1997, L. A. Inkster). (There are three old records of the species at aggregate level from other parts of the county, but further details of these are not available.)

*__R. × sterilis__ (_R. nasturtium-aquaticum_ × _R. microphylla_). _Hybrid Water-cress._ Casual. In a small watercourse by the shop at Swarthoull, Hillswick, Northmavine (prior to 1993, R. C. Palmer and WS), and said to have originated from 'an island near

Glasgow'; a large patch by a rill on a low sea-bank just S of 'Burrastow House', West Mainland (1993, WS), planted *c.*1991 and of Swarthoull origin. In 2008 WS could not find it at either place.

*R. microphylla. *Narrow-fruited Water-cress.* A very rare and well-established alien known from one site: in the NE corner of the Loch of Forratwatt, near Walls, West Mainland, in shallow swampy conditions. Origin unknown but probably planted. It has almost certainly been here since at least 1968 when it was found by R. C. Palmer.

*R. palustris. *Marsh Yellow-cress.* Casual. First recorded, in 1924, at the Olna Firth whaling station, North Mainland. There are two comparatively recent records: one plant within the grounds of the Lerwick Generating Station (1985), and one in a pipe-yard at the Norscot Base. Lerwick (1996, WS).

*Armoracia rusticana. *Horse-radish.* Casual. A mere garden escape or outcast.

Cardamine pratensis. *Cuckooflower.* One of our common species of damp or wet grassland, by burns and ditches, etc. Double-flowered forms are sometimes recorded.

*C. flexuosa. *Wavy Bitter-cress.* A well-established alien in and around Kergord Plantations, Weisdale, Central Mainland. Otherwise an occasional garden weed (often under trees) which, as an escape or outcast, occurs by grassy burns (sometimes among irises), by ditches, and on dumped garden soil, etc. In 2008 WS found several patches, some quite large, in a grassy field above the W side of the head of the East Voe of Scalloway, in areas where animals had been fed during the winter.

C. hirsuta. *Hairy Bitter-cress.* As a native this is found occasionally and sparingly throughout Shetland in mossy-grassy turf in rocky places, as on the W side of the Clift Hills, South Mainland (sea-banks, Forsan (HU3826), 1998, WS), rarely on rocky holms in lochs. Hairy bitter-cress is otherwise a well-established alien, especially in gardens where it can become a nearly ineradicable weed.

Arabis petraea. *Northern Rock-cress.* A rare arctic-alpine of debris, rocks, and steep sea-cliffs on the serpentinite formations of Unst (Keen of Hamar, and the Hill of Clibberswick sea-cliffs) and Fetlar (N of the Loch of Winyadepla, and the East Neap sea-cliffs).

*A. hirsuta. *Hairy Rock-cress.* Casual. Once found (in 1928) as a garden weed at Brough Lodge, Fetlar.

*Lunaria annua. *Honesty.* A casual sometimes seen as a garden outcast.

*Lobularia maritima. *Sweet Alison.* The above remark also applies here.

Draba incana. *Hoary Whitlowgrass.* An arctic-alpine of dry rocky outcrops and associated grassy niches on limestone, serpentinite, and (occasionally) granite. Local, rarely in quantity, and deteriorating in some sites because of the vulnerability of its habitat to the combined effects of overgrazing and trampling by sheep. In recent years a number of new stations have come to light. The following four are all in Central Mainland: N of Utnabrake, near Scalloway (1996, 1998, L. A. Inkster); in the Cova/Haggersta area of Weisdale (1995, LAI); just N of Jackville, Strom Ness peninsula (1993, WS); W side of the Strom Ness peninsula, about opposite the township of Stromness (1992, LAI). The following two are in West Mainland: at least twenty plants on a bare rock sitting in the NW corner of the Loch of Vaara (2003, WS); by or near both sides of the loch at Ness, Sandness (1988, D. H. Dalby). About twenty plants on a hillock NE of Seggi Bight, Vementry (1999, LAI). About 375 m E of the N end of the Loch of Snarravoe, Unst (2001, LAI and WS); Setters Quarry,

Unst, by the road between Baltasound and Haroldswick, fairly sparingly within the surrounding fence (1995, LAI).

Cochlearia officinalis. *Common Scurvygrass.* One of our commonest seashore plants and one which has caused taxonomic problems for a long time. These seem set to be with us for a while. Stace (1997) recognises three main taxa within the complex: *C. officinalis*, *C. pyrenaica*, and *C. micacea.* The last two are on record from Shetland but are not now accepted for the county. This leaves *C. officinalis* sensu stricto within which exist two fairly distinct subspecies. Subsp. **officinalis** is a large, fleshy-leaved, apparently always white-flowered coastal plant which is at its best on guano-enriched cliffs, stacks and skerries, and on beaches with rotting seaweed. It is also found on a very few holms in lochs. Subsp. **scotica**, on the other hand, is a very small prostrate taxon with mauve (sometimes white) flowers and occurs commonly in fine peaty, sandy, or muddy turf by the coast. A similar plant grows on the serpentinite tracts of Unst and Fetlar, either on debris or in turf among outcropping rocks. A long-podded variant is also found here. This gave rise to the records of *C. pyrenaica* and *C. micacea*, as a result of too much dependence being placed on one unreliable character. The status of these serpentinite plants still remains problematical.

C. officinalis sensu stricto × **C. danica**. In 1979 this was seen on an old wharf at the North Ness, Lerwick. This hybrid may occur more widely; robust plants with lilac-flushed flowers should be carefully examined.

C. danica. *Danish Scurvygrass.* Sea-cliffs, stacks, and coastal stonework (notably brochs), rarely on beaches. Occasional in South and Central Mainland, rare or unrecorded elsewhere. However, it does extend to Yell and Unst (sparingly in both), and occurs in Fair Isle, Colsay, Papa Stour, etc.

*****Camelina sativa**. *Gold-of-pleasure.* Casual. Twice recorded, in 1967 and 1968, on the Lerwick rubbish-dump (on the site of the North Loch).

*****Capsella bursa-pastoris**. *Shepherd's-purse.* A well-established alien which is frequent to common in waste rubbly places, by roadsides, and in arable ground (especially on sandy soils).

*****Thlaspi arvense**. *Field Penny-cress.* A rare casual of gardens and (rarely) waste ground. There are three new records: garden of 9 Knab Road, Lerwick (1998, I. Clark); Fair Isle Bird Observatory garden (1999, N. Riddiford); 'Braknahool', Scalloway, on builders' rubble (2006, WS).

*****Iberis umbellata**. *Garden Candytuft.* Casual. Merely a garden outcast.

*****Lepidium sativum**. *Garden Cress.* A rare casual of waste ground. There are no recent records.

Subularia aquatica. *Awlwort.* A rather small, apparently rare, and easily overlooked aquatic. It is found on the muddy-stony bottoms of a few lochs in West Mainland, Northmavine, and in Papa Stour. In West Mainland it grows in the Lochs of Kirkigarth and Bardister; Burga Water; two lochs between Burga Water and the Loch of Whitebrigs; several scores of plants in a small bay by the N side of Hulma Water, and one or two plants in the NE corner of the loch (2003, WS); the most westerly of the Smalla Waters, near Sulma Water (2006, N. Aspey). There was 'quite a lot' in the Dutch Loch, Papa Stour (2008, British Bryological Society field meeting). In Northmavine it occurs in two of the Many Crooks group of lochs in North Roe. Awlwort is usually seen in very small numbers.

Diplotaxis tenuifolia. *Perennial Wall-rocket.* Casual. Once recorded, in 1961, in a garden in Breiwick Road, Lerwick.

*Brassica oleracea var. **viridis**. *Kale.* Casual. Formerly a frequent outcast or relic of cultivation by burns and on rubbishy foreshores in the crofting districts. Kale is much less grown nowadays.

*B. **napus** subsp. **oleifera**. *Oil-seed Rape.* In 1998 and 1999 this was a frequent casual on rough ground at Blacksness Pier, Scalloway (WS). It was never seen again.

*B. **napus** subsp. **rapifera**. *Swede.* Casual. A relic or outcast from cultivation.

*B. **rapa** subsp. **campestris**. *Wild Turnip.* A casual of recently sown roadside verges, with only two localised records: by the new road above the Bight of Vatsland, N of Lerwick (1985, WS); between Westshore and Port Arthur, Scalloway (1992, WS).

*B. **rapa** subsp. **rapa**. *Turnip.* Casual. A relic or outcast from cultivation.

*B. **juncea**. *Chinese Mustard.* A very rare casual, twice recorded: one plant on shingle at the head of the Voe of Sound, Lerwick (1999, R. C. Palmer); frequent on the Fair Isle Bird Observatory seed dump (1999, N. Riddiford and F. H. Perring).

*B. **nigra**. *Black Mustard.* A very rare casual which was first found, in 1969, on waste ground, Chromate Lane, Lerwick. There are two recent records: garden of 9 Knab Road, Lerwick (1996, 1998, I. Clark); Fair Isle Bird Observatory seed dump (1999, N. Riddiford and F. H. Perring).

*Sinapis arvensis. *Charlock.* This well-established alien was formerly frequent in crofting and farming areas in arable ground and around farm buildings, on waste ground, etc., on the better soils. It is still widespread but more thinly distributed owing to loss of habitat.

*S. **alba** subsp. **alba**. *White Mustard.* Casual. Very rare and with no records for a long time. First noted, in 1890, in cultivated fields at Walls, West Mainland, where it was seen again, in 1968, on a disturbed wayside. In 1953 it grew commonly on the Lerwick rubbish-dump (by the Loch of Clickimin) and was seen twelve years later on the replacement tip (on the site of the North Loch). (There is a pre-1970 record for HU45 but the writer has been unable to discover any further details.)

Cakile maritima. *Sea Rocket.* Frequent by the coast on sandy or shingly beaches and on the seaward edge of sand-dunes, often forming large permanent colonies on the larger beaches, Nor Wick, Unst, being a good example. Seedlings and small plants also appear by the coast but are transient.

*Rapistrum rugosum. *Bastard Cabbage.* A very rare casual. Two plants in 1977 on waste ground near 'The Studio', New Street, Scalloway.

[**Crambe maritima**. *Sea-kale.* Believed to be extinct. Around 1991 T. Angus found one flowering plant on a boulder beach at the Taing of Helliness, Cunningsburgh, South Mainland, almost certainly from a seaborne seed. The local sheep ensured that it did not appear again. Sea-kale is very likely to be found again somewhere in Shetland, but perhaps not for a long time.]

*Raphanus raphanistrum subsp. **raphanistrum**. *Wild Radish.* A well-established alien of arable ground (notably cornfields), around crofts and farms, and on waste ground, etc. Less frequently seen nowadays owing to loss of habitat through changing agricultural practices.

Empetrum nigrum subsp. **nigrum**. *Crowberry.* A common shrub on heaths and moors (from deep wet peat to dry stony fellfield), and on steep banks and cliffs by the coast where it often forms large conspicuous patches; exceptionally in peaty turf on the walls of ruined crofthouses, as at Houllmastouri, E side of Bressay (2000, WS).

E. nigrum subsp. **hermaphroditum**. An apparently very scarce plant. Sparingly among boulders on the N side of Ronas Hill, Northmavine. The first definite record of this was made in 1987 when R. W. M. Corner found it in HU3084 at an altitude of 250–300 metres. This was followed by two further sightings in the same general area: N of the loch at the Shurgie Scord (the highest loch in Shetland) at *c*.350 m (1990, WS); same altitude but NW of the loch, and also farther west, above the source of the Burn of Black Butten, same altitude (2003, P. V. Harvey and WS). There must be other stations on Ronas Hill, apart from those just listed for HU3084 and HU3184, but searching for them could be tiresome owing to the large area involved and the relative abundance of the common subspecies. (The reference to HU3085 in Scott, *et al.* (2002) is an error for HU3184.)

Loiseleuria procumbens. *Trailing Azalea.* Very rare but not uncommon within the wide limits of its single station. This arctic-alpine shrub grows on bare granite fellfield and associated *Rhacomitrium*-heath on Ronas Hill, Mid Field, and Roga Field, all in Northmavine. On the NE sides of the last two it descends to 240 metres.

*Gaultheria mucronata. *Prickly Heath.* Very rarely planted in semi-open places.

Arctostaphylos uva-ursi. *Bearberry.* Generally a rare shrub of dry, stony or rocky *Rhacomitrium*-heaths but locally frequent in parts of Northmavine, as follows: between Mangaster Voe and Gunnister Voe; between Glussdale Water and Punds Water to the N; on the Hill of Burriesness, entrance to Ronas Voe; and scattered over a large area of North Roe from near the Moshella Lochs to the Beorgs of Housetter N to the Inniscord Lochs and the Beorgs of Skelberry. Bearberry also grows in Muckle Roe from where it has been known for over two centuries.

A. **alpinus**. *Arctic Bearberry.* This arctic-alpine shrub, rarer than the preceding, is found in the same general habitat (sometimes a little damper and peatier) than that favoured by its more frequent relative, the two not uncommonly growing together. Arctic bearberry is found in two widely separated areas in Northmavine: the W and N slopes of Ronas Hill, and in an area centred on Egga Field, between the Sae Waters and Brettoo Loch. An old record from Yell, 'a very large plant on the undercliff ', has never been confirmed.

Calluna vulgaris. *Heather.* An abundant shrub over much of upland Shetland, and dominant over large tracts of land, especially in Yell and in numerous other places. While preferring dry ground, it is not averse to growing in boggy sphagnous places.

Erica tetralix. *Cross-leaved Heath.* A common shrub on boggy moors in areas of deep peat and on wet heaths over poorly drained shallow soils.

E. cinerea. *Bell Heather.* Another common shrub on dry moors, mossy or stony heaths, on sunny, craggy rocks, and by rocky-heathery burnsides; also in peaty, boggy ground and on wet heaths and then often accompanied by *E. tetralix.*

Vaccinium oxycoccos sensu lato. *Cranberry.* A species of cranberry was recorded in 1985 from Hascosay, between Yell and Fetlar. Since then many attempts to relocate it have been made; all have failed. It may have been *V. microcarpum* (small cranberry), the species recorded from Caithness. There are no cranberry records from Orkney or Faeroe. Small cranberry grows in miry, sphagnous places in mainland Scotland, a habitat not at all unusual in Hascosay, or in Shetland generally.

V. vitis-idaea. *Cowberry.* This is essentially an evergreen, rarely flowering shrub of heaths and moors (*Rhacomitrium*-heaths, edges of peat-haggs, etc.) of the upper slopes and summits of many of our higher hills from Fitful Head (South Mainland) to Saxa Vord (Unst) and from the Sandness Hill range (West Mainland) to the Ward of Bressay. It also occurs on the North Roe plateau, Northmavine, where it is rather sparsely distributed over a large area and is only occasionally locally common. Cowberry descends to *c.*100 m. on the rocky banks of the stream from Crookna Water, Heylor, Northmavine.

V. uliginosum. *Bog Bilberry.* This local shrub favours moorland slopes and summits and *Rhacomitrium*-heaths at high altitudes. It can be found on two hills in Foula; Royl Field (South Mainland); the Sandness Hill range, West Mainland; on three summits between Weisdale and Voe, Central Mainland; Hill of Dale (HU4069), N of Voe, North Mainland (1994, WS). In Northmavine it grows 200 m NE of the NE corner of Lergeo Water, in the Tingon district (2002, B. O'Hanrahan); on Ronas Hill, Roga Field, and Mid Field; and on the northern holm in the loch at Clubbi Shuns, North Roe. The only recorded station in Yell is on the Ward of Otterswick, but in Unst it can be seen on Hermaness Hill, Sothers Field, and Saxa Vord. The Tingon site, at ninety-six m, is the lowest recorded for Shetland.

V. myrtillus. *Bilberry.* A frequent and often dwarfed shrub on well-drained moors, *Rhacomitrium*-heaths, steep-sided sheltered streamsides and holms in lochs where (in these two habitats) it may become much bushier. Bilberry prefers the higher ground and is often a regular component of the moorland flora over large areas.

[**Pyrola** sp. *Wintergreen.* Believed to be extinct. There are three records for the genus. The first was made in 1769 by J. Robertson who claimed it for Shetland but provided no further details. The second was by C. Fothergill, in 1806, who recorded *P. rotundifolia* from the 'south side of Ronas hill', Northmavine. Unfortunately, the record is unreliable as Fothergill has, elsewhere in his list in connection with another species, mixed up Orkney and Shetland, and we know that he had seen *P. rotundifolia* in Orkney in the same year. The third record is rather more reliable, 'near Walls', West Mainland (Edmondston 1845), under *P. media*. H. C. Watson, a noted British botanist at the time, said that Edmondston had earlier recorded his plant under *P. rotundifolia* but the evidence for this cannot be found. Perhaps Edmondston's Walls plant was *P. rotundifolia*: after all, it is this species which grows in Orkney.]

Primula vulgaris. *Primrose.* One of our most popular native plants. Frequent and sometimes common on coastal banks, cliffs and pastures, and in the limestone valleys of Central Mainland. Primroses have been dug up, unnecessarily, for the adornment of gardens, and have also been introduced to areas where they did not originally occur.

*****Lysimachia punctata**. *Dotted Loosetrife.* Casual. Often grown in gardens and thus occasionally seen as an outcast near houses or where it may have been planted.

Trientalis europaea. *Chickweed-wintergreen.* A local plant of fine heathy ground, often near the sea; rarely at higher altitudes. In 1998 T. Russell counted well over 1,000 plants on the NW slopes of Compass Head, near Sumburgh, South Mainland, an old station.

Anagallis tenella. *Bog Pimpernel.* An attractive local plant of damp turf by lochs and burns, and in sphagnous bogs and marshes. It has a distinct preference for South and West Mainland, and also occurs in Fair Isle, Foula, Vaila, Bressay (a little way inland from Grut Wick, 1987, M. Robertson; in large numbers around Sand Vatn, 1991, P. Newton), Out Skerries, Fetlar (Loch of Stivla area, 1991, M. Tickner), and Unst. There are no records from Whalsay, and from Yell there is one doubtful report. Flint (2008) records a double-flowered form from Housay, Out Skerries, an apparently very rare variation.

A. arvensis subsp. **arvensis**. *Scarlet Pimpernel.* Casual. An occasional garden weed. The writer and R. C. Palmer were told that prior to 1960 a blue-flowered pimpernel had appeared in a Scalloway garden. However, as it was never seen again it is impossible to tell whether it was subsp. *foemina* (blue pimpernel) or the blue form of scarlet pimpernel which, not surprisingly, is easily mistaken for it.

Glaux maritima. *Sea-milkwort.* A common coastal plant of saltmarshy turf, quiet muddy beaches, on rocks well within the spray zone, and in barish cliff-top sward and on open gravelly ground by the coast.

***Ribes rubrum**. *Red Currant.* Casual. Very rarely seen in even semi-wild situations and, as Scott & Palmer (1987) comment, 'scarcely worthy of inclusion'.

***R. nigrum**. *Black Currant.* Casual. Sometimes planted by burns, as by the Burn of Laxdale, Cunningsburgh, South Mainland (1999, R. C. Palmer), very rarely an outcast.

***R. sanguineum**. *Flowering Currant.* Casual. This is a popular garden hedge in some parts of Shetland. It is occasionally seen as an outcast or where it has been planted by burns in the inhabited areas, and in other places not far from houses.

***R. uva-crispa**. *Gooseberry.* Casual. Sometimes planted by burns; otherwise rarely seen by old walls or as a garden relic.

Sedum rosea. *Roseroot.* A frequent arctic-alpine in many rocky places by the coast, especially on high, steep sea-cliffs in the W and N of Shetland. It is rarely seen far from the coast, as on crags by the SW side of Ronas Hill and on a holm in Sandy Water, both in Northmavine. Often grown in gardens and sometimes seen an an outcast in their vicinity.

***S. telephium**. *Orpine.* Casual. This garden plant is sometimes seen as an outcast or deliberate introduction.

***S. spurium**. *Caucasian-stonecrop.* Casual. The above remarks also apply here.

***S. acre**. *Biting Stonecrop.* Casual, rarely planted outside enclosures. This garden stonecrop was introduced on limestone outcrops many years ago at Mousa Ness, Whiteness (last seen in 1993) and at Brough, South Nesting (last seen in 1984). In both of these Central Mainland locations it survived for a long time; in one or two others it soon vanished.

S. anglicum. *English Stonecrop.* Dry rocky-heathy grassland and rocky outcrops on or near the coast in only two places in the county. Very rare, but not uncommon in some parts of Out Skerries where it extends to the summit of the Ward of Bruray. It is the 'beautiful white flower' which C. Fothergill in 1806 noted on Bruray, the first record of it from Shetland. The other station is at Ling Ness, South Nesting, Central Mainland, where it is very scarce. It has also occurred, in 1993, as a somewhat frequent casual on levelled ground and verges by the minor road leading to the port complex at Sella Ness, near the Sullom Voe Oil Terminal, North Mainland, and was probably introduced during the 1970s when the harbour works were being built. By 2000 it had become scarce, and in 2008 WS saw it only in front of the premises of C. & R. Diving Ltd, on mossy-gravelly ground. Its flowers were of a size typical of the species, unlike the smaller blooms which sparingly adorned the plant when it formerly grew more widely in the area, but under rather starved conditions.

*Darmera peltata. *Indian-rhubarb.* Casual. Noted in 1975 as a scarce garden relic, 'Leagarth House', Houbie, Fetlar. It was present in 1982 but was not seen again.

*Saxifraga × urbium (*S. umbrosa × S. spathularis*). *Londonpride.* Casual. This is a common garden plant which is often seen as an outcast or deliberate introduction.

S. oppositifolia. *Purple Saxifrage.* This, an arctic-alpine plant and our only native saxifrage, is found in two main areas, Northmavine and West Mainland. In the first-named division it occurs in many places by the eastern seaboard (on sea-cliffs and wet gravelly coastal slopes) from the N side of the Ness of Queyfirth (one of its best sites in Shetland) to the Wick of Virdibreck, E of Isbister; on the N-facing coast from near Hevda and E to the shore below the Hill of Breibister, near Fethaland; and at Roer Mill, NNW of Sandvoe. It can also be found in a few places some way inland in the same parish: by the track N of Benigarth; near the main road at Claypow; and sparingly *c.*300 m N of Ure Water. In the West Mainland division a few plants can be found on the Snarra Ness peninsula, on rocks near a small loch.

*'S. × arendsii hort.' *Mossy Saxifrage.* Casual. A convenient but invalid label for a group of very variable hybrids of perhaps more than two species, and which are commonly grown in local gardens. In 1990 several scores of tufts or clumps grew on the seaward embankment of the A970, directly opposite the Shetland Hotel, Lerwick, where it was being grown at the time.

Parnassia palustris. *Grass-of-Parnassus.* Base-rich marshy places, stony flushes by lochs, and (formerly) in damp dune-pastures. Local to frequent in West and Central Mainland, local in North Mainland, and very scarce in Northmavine and South Mainland. There are two records from Unst: Baltasound (early 1920s); W side of Uyea Sound (1962). No later sightings from either place are known. It is unrecorded from Whalsay, Yell, and Fetlar. An attractive large-flowered form (var. *condensata*) used to grow abundantly on the formerly unspoiled links between the N end of the Loch of Spiggie and the sand-dune fringe behind the Bay of Scousburgh, South Mainland. Unfortunately, agricultural activities soon destroyed the best station in Shetland for this fine plant. By 1999 only a few small specimens could be found.

*Spiraea sp. *Bridewort.* Commonly grown in gardens and occasionally seen outside them (sometimes by burns) as an outcast, straggler, or deliberate introduction. This is a somewhat taxonomically difficult groups of shrubs. More than one species may be present in Shetland, and the placing by Scott & Palmer (1987) of all local material under *S. salicifolia* may have been wrong.

Filipendula ulmaria. *Meadowsweet.* A frequent plant of the better soils, especially limestone. Wet or marshy places (but not permanently waterlogged), often by lochs (notably by the NW side of the Loch of Tingwall, Central Mainland), holms in lochs (particularly the islands in the Lochs of Bardister and Breibister, near Walls, West Mainland), and occasionally in damp rocky pastures.

Rubus saxatilis. *Stone Bramble.* A local to frequent plant of basic soils, especially limestone. Rocky places: streamsides, outcrops, and on sea-banks (particularly in West Mainland). Also on the serpentinite fellfield in Unst, and sparingly, and not recently, on the granite fellfield of Ronas Hill, Northmavine (on the summit, one plant, 1958, WS). A large patch occurs by the roadside just N of Breck, Whiteness, Central

Mainland. Stone bramble flowers are rather sparingly produced but, in some seasons at least, it flowers well on the Keen of Hamar serpentinite fellfield, followed very occasionally by a scarlet fruit.

R. idaeus. *Raspberry*. Casual. This is merely a straggler or outcast from gardens, and sometimes a deliberate introduction.

R. spectabilis. *Salmonberry*. A well-established alien which is frequently seen as a garden outcast or deliberate introduction: burnsides, roadsides, etc., as at the Burn of Tresta, Central Mainland, where it has been known for many decades.

R. lindebergii. *Bramble*. A well-established alien in one place: by the Burn of Laxdale, Cunningsburgh, South Mainland, below the Tow–Gord lane. First noticed, in 1978, by R. C. Palmer. Still there in 2008 (WS).

R. echinatoides. *Bramble*. Casual. This or a close relative was common, at least in 1976, in Leegarth Plantation, Kergord, Weisdale, Central Mainland, but by 1982 it had disappeared.

R. latifolius. *Bramble*. A well-established alien by the stream at Linds, Sandwick, South Mainland. It was running wild when it was first noticed in 1962 and is still there, having survived not only attempts to restrict its spread but also the severe flooding after a rainstorm in 2003 which removed some of the bramble. It, or a taxon very close to it, grows in Peggy's Plantation, Kergord, Weisdale, Central Mainland, where it has been known for many years.

R. pictorum. *Bramble*. A well-established alien. [Planted about the mid 1990s] by the Sandsound road, Central Mainland [on the bend WNW of Kirka Taings] (1999, R. C. Palmer). The patch was nine m long in 2008 (WS).

Potentilla palustris. *Marsh Cinquefoil*. A common species of marshy or swampy places, often dominant over large areas; also in ditches, by burns, and on holms in lochs.

P. anserina. *Silverweed*. In Shetland this is essentially a species of sandy or shingly beaches and coastal pastures, its long trailing stolons being a feature of the upper part of many beaches. It can also grow well inland, though much less commonly, in damp pastures, marshes, by roadsides, and on waste ground.

P. erecta subsp. **erecta**. *Tormentil*. A very common plant of peaty soils from bogs and deep sphagnous moors to dry shallow heaths. It varies greatly in size from the small plants which begin to cover recently burned moorland to the much larger examples which straggle through the mixed herbage of well-vegetated peaty holms in lochs. Sir Walter Scott, in 1814, noted it on the moors near Lerwick.

*Fragaria × ananassa. *Garden Strawberry*. Casual. Merely a rare garden outcast or (perhaps once) a bird-sown introduction.

Geum rivale. *Water Avens*. A local plant of rich burnsides, rocky loch margins, damp rocky grassland, and (rarely) in damp pastures, on the various limestone bands of central Shetland from near Utnabrake, near Scalloway, Central Mainland, N to the Burn of Sandgarth, above the head of Dales Voe, Delting, North Mainland. The Burn of Sandgarth record (1950s, B. Johnson) has only recently come to light; the plant was still there (2006). Outside of the main limestone areas there are a number of records: Gulberwick, near Lerwick, Central Mainland (1979); Norby (1980), Mousavord Loch (1982), Kellister (1964), and Effirth (1999), these four in West Mainland; Braeside (*c.*1958) and Eastwick (1966), both at Ollaberry, North Mainland; Fetlar (1997); and the Loch of Watlee (1981) and Muckle Heog (2004), both in Unst. The bracketed dates are the latest sightings; in some of these stations the plant may no longer be found.

*G. sp. or cultivar. *Avens*. Casual. A rare garden outcast or, perhaps, an occasional deliberate introduction.

*Acaena anserinifolia. *Bronze Pirri-pirri-bur*. A rare casual or garden outcast. Two records: a small patch by the A970 near Scalloway, 1958, destroyed the following year by digging; a small patch by the road at the Mossbank Primary School (now the Old School Centre), 1991 (WS), still there in 1997 but not found in 2005 (WS).

Alchemilla alpina. *Alpine Lady's-mantle*. A very rare arctic-alpine plant which, however, is not at all uncommon above *c*.275 m on bare granite fellfield on the summits and slopes of Ronas Hill and neighbouring Mid Field, Northmavine, its only station in the county. Occasionally a plant or two may be seen much lower down (*c*.150 m) by heathery streamsides (as by the Grud Burn) and, at a higher altitude, among rocks at the top of the Stonga Banks. These sheltered plants are much larger than those on the exposed fellfield.

*A. conjuncta. *Silver Lady's-mantle*. Casual. In 1898 this grew in the lawn and flower-beds of the (then) Queen's Hotel, Baltasound, Unst. In 1920 and 1921 it was still there with a plant or two in an adjoining pasture. No later records from here are known to the writer. In Fetlar, in 1958, it was overrunning the remains of the garden at 'Leagarth House', Houbie, but by 1975 it had declined to a small patch or two. It stayed at this level for many years until 2008 when the remainder were transplanted to pots by a neighbouring gardener.

A. filicaulis subsp. **filicaulis**. *Lady's-mantle*. Grassy rocky outcrops and streamsides, dry pastures, etc., on limestone and other basic soils. Local to frequent on the various limestone bands of Central Mainland, rare or local elsewhere, and unrecorded or absent from all the islands except Mainland. In its main habitat (grassy niches among rocks) severe grazing and trampling by sheep have eliminated it or reduced it to tiny flowerless plants in several areas.

*A. glabra. *Lady's-mantle*. A well-established alien of grassy burnsides, roadsides, old yards, etc., rarely far from houses, and probably always of garden origin. First noted in Shetland, in 1911, at Baltasound, Unst. In 1991 WS found a few tiny plants in the turf of the embankment of the dam at Helliers Water, Unst.

*A. mollis. *Lady's-mantle*. Casual. An invasive garden plant sometimes seen as an outcast or deliberate introduction by roadsides in or at some distance from the inhabited areas. This prolific seeder was first recorded, in 1989, on gravel at Blacksness Pier, Scalloway, clearly from a flower-border near Scalloway Castle (WS). In 2008 there were many plants in this area.

*Aphanes arvensis sensu lato. *Parsley-piert*. Perhaps a rather scarce and weakly established alien long ago; in recent dcades, however, it has become a very rare casual and was last seen, in 1982, at Burland (Trondra) and Aith (Fetlar), among potatoes in both cases. Grassy rocky outcrops; dry pastures; and cultivated ground. It was found mainly on the better soils of South and Central Mainland, and in Northmavine, Unst, Yell, and Fetlar. The great reduction in the cultivation of oats, potatoes, etc., has undoubtedly led to its present rarity; nevertheless, it is unlikely to be extinct. The only segregate definitely recorded from Shetland is **A. australis** (slender parsley-piert).

30

***Rosa rugosa**. *Japanese Rose*. A well-established alien and a very popular and hardy rose of gardens which often appears outside enclosures (often by burns) as an outcast, straggler, or deliberate introduction, as on the holm in the Loch of North-house, N of Twatt, West Mainland, where it was planted years ago and has completely covered the small island. Japanese rose usually produces fruits abundantly and conspicuously in early autumn.

*__R__. 'Hollandica'. *Dutch Rose*. Casual. A hybrid involving *R. rugosa,* rarely seen in or outside gardens: planted by the Ham Burn, Foula, before 1959 but almost dead in 1984; a garden relic (in 1980) at 'Engamoor', West Burrafirth, West Mainland.

R. **caesia** subsp. **glauca**. *Glaucous Dog-rose*. Now almost restricted to inaccessible places: grassy or heathery coastal cliffs and banks; steep-sided burns; outcrops; and holms in lochs. It is most frequent by and near the S and E shores of St Magnus Bay. Elsewhere it occurs locally and in small quantity, but is not uncommon by the W side of Whale Firth, Yell.

*__R__. **sherardii**. *Sherard's Downy-rose*. Casual. Planted sometime after the mid1950s outside the wall of the (then) South Nesting Primary School, Vassa, Central Mainland.

*__R__. **rubiginosa**. *Sweet-briar*. Casual. Planted before 1965 by the Ham Burn, Foula. It was still there in 1975 but by 1982 it had died out.

Malus sylvestris. *Crab Apple*. This very rare shrub was added to the Shetland list in 1991 by L. A. Inkster and WS who found about six patches on a steep rocky sea-bank at the Neap of Foraness, Dales Voe, Delting, North Mainland, growing with eared willow, glaucous dog-rose, and honeysuckle. Another site was discovered by them the following year, one patch (apparently) in a *geo* on the N side of Quey Firth, Northmavine, with no accompanying shrubs. It was last seen at this site in 1993; sadly, by 2000 it and the rocky face on which it grew had fallen into the sea through undermining of the site by heavy seas. Fortunately, the Neap of Foraness station looks very secure. Crab apple has not been seen in flower at either site.

*__M__. **domestica**. *Apple*. Casual. Perhaps a rare garden outcast, but more usually seen where apple pips have been discarded. Tiny saplings occur from time to time by roadsides and ditches, on rubbly ground and cattle-grid floors, etc. Very few of them survive to become small bushes which may persist for several years. One such bush, at Levenwick, South Mainland, lasted for at least forty-eight years before being dug up to make way for a passing-place.

Sorbus aucuparia. *Rowan*. Crags, holms in lochs, rocky burnsides, and (rarely) on sea-banks. Local in the N part of West Mainland, but frequent in Northmavine, as in the Ronas Voe and Ronas Hill areas, and on the North Roe plateau. Very rare elsewhere in Shetland. No modern records exist for Fetlar or Unst, and the single recent record from Yell needs confirmation. One of the best examples in Shetland is the bush on the holm in the North Loch of Hostigates, West Mainland. Extremely tiny plants (two cm across and with a minute leaf or two) have been seen in very small numbers over several decades on dry rocky streambanks in the Heylor and Assater areas of Northmavine. These seem unable to grow any larger and are best regarded as short-lived plants which have sprung from bird-introduced seeds from time to time. They may well have originated in the nearby Ronas Hill area, or from local gardens. Some garden rowans are themselves of native Shetland origin, having been taken into cultivation from the wild.

***S. aria** sensu lato. *Common Whitebeam.* Casual. In 1969 a bush, probably from bird-sown seeds, grew on the wall of a ruined house, Burns Lane, Lerwick. The house was removed or rebuilt some years later. Otherwise rarely seen outside enclosures.

***S. latifolia** sensu lato. *Broad-leaved Whitebeam.* Casual. In 1973 a sturdy sapling, probably from bird-sown seeds, grew on the wall of a ruined house, Reform Lane, Lerwick, but had vanished two years later.

***Crataegus monogyna.** *Hawthorn.* Casual. Very rarely planted outside enclosures. Long-persistent in two places: by the lane above 'Gardie House', Bressay (planted *c.*1810); between Veensgarth and Gott, Tingwall (planted prior to 1841), now very few left but one has been a small tree (five m high) for many years following shelter obtained from a nearby building.

Anthyllis vulneraria. *Kidney Vetch.* A local to frequent plant of dry, sunny, grassy coastal cliffs and banks and short maritime turf. Also some way inland as on the serpentinite of Unst (fellfield) and Fetlar (debris). Less often elsewhere inland in rich grassland, by roadsides, etc. Mostly, if not entirely, represented by subsp. **lapponica**. For a discussion on the colour forms found in Shetland see Scott & Palmer (1987). Alien strains have occurred by roadsides, most noticebly where the plant had not grown before, or where the habitat is clearly unsuitable.

Lotus corniculatus. *Common Bird's-foot-trefoil.* Frequent to common on dry grassy coastal banks, cliffs and (often sandy) pastures, and inland on rocks, stony heaths, etc. Imported alien strains in seed mixtures almost certainly occur by roadsides.

***L. pedunculatus.** *Greater Bird's-foot-trefoil.* Apparently a relative newcomer to Shetland and now a well-established alien in at least two places. It was first noticed in a field near the Loch of Clickimin, Lerwick, in 1969, 'in vast quantity', but by 1983 it had been destroyed by the building of the Clickimin Leisure Complex. Also in 1969 it grew in a cul-de-sac by a house in South Road, Lerwick; by 1980 it had vanished. A third site was discovered, probably in 1984, by D. H. Dalby: in abundance in a marshy area and by nearby roadsides at the Hillswick cemetery (at the neck of the Ness of Hillswick, Northmavine). In 1988 he found it in even greater quantity on grassy streambanks and roadsides at the head of East Burra Firth, Central Mainland. It still flourishes in Dalby's two stations, but was less abundant than in recent years below the road at his latter site (2008, WS). The latest records go back to 1997 when it was recorded by the Burn of Hoswick and to the W of the Burn of Setter, both at Sandwick, South Mainland, by Carrick *et al.* (1997).

Vicia cracca. *Tufted Vetch.* Local to frequent in or near cultivated ground, on dune-pastures (often profusely), grassy sea-banks, and rocky serpentinite heaths; rarely by rocky burnsides and on holms in lochs. It is also found on a handful of small offshore islands, luxuriant where ungrazed, as on the Skerry of Eshaness, off Northmavine (2003, WS), but tiny and flowerless, possibly as a result of grazing, on a few others.

***V. benghalensis.** *Purple Vetch.* A very rare bird-seed casual, once recorded: frequent in 1994 in the garden of 9 Knab Road, Lerwick.

***V. hirsuta.** *Hairy Tare.* Another very rare casual and one which has not been seen for a long time. Field W of Walls, West Mainland (1954); arable ground, Skellister, South Nesting, Central Mainland (1960); near the pier, Bruray, Out Skeries (1961).

***V. tetrasperma.** *Smooth Tare.* Casual. Very rare. Foula, only one plant, as a field weed at Brae, above Ham Voe (1999, S. Gear).

V. sepium. *Bush Vetch*. Almost certainly native in some stations at least: holms in the Loch of Clousta (1901, 1960, 1982/3); and on a holm in the N end of Burga Water (1902, 1958), both lochs in West Mainland; rocky burnside, Burn of Valayre, near Brae, North Mainland (1959, 1973, 1991, 2008); and occasional patches among *Luzula sylvatica* on steep sea-banks on the E side of Fora Ness, near Sand, West Mainland (1987, WS). Otherwise local to frequent as a possible alien in arable ground and rough roadside herbage.

*V. sativa sensu lato. *Common Vetch.* A very scarce casual which was probably more frequent long ago when cornfields (its chief habitat) were more numerous. It has had a long connection with Unst: in 1886 it was a cornfield weed in the island; in 1911 it grew in the Baltasound area; and in 1950 it was a weed in fields at Norwick. Recently recorded from Foula (Gear 2008). Apart from arable ground it has been found in rubbly places and, in 1990, on a reseeded roadside embankment by the Yell Leisure Centre (A. D. D. Gear). Subsp. **sativa** is the only subspecies definitely on record from Shetland.

*V. faba. *Broad Bean.* Casual. Two records from Lerwick: roadside embankment by the Loch of Clickimin, several plants (1983); rubbly ground, Holmsgarth Ferry Terminal, many plants (1985).

Lathyrus japonicus. *Sea Pea.* Probably nearing extinction if still present. This was first recorded, in 1837, from the sands at the head of Burra Firth, Unst. It died out here shortly after 1900 following a period of steady deterioration. In 1959 at nearby Nor Wick a new station was discovered among the dunes behind the beach. Here, it flourished for a while—and was an attractive addition to the local flora—but it, too, declined for no apparent reason, and by 2001 only one flowerless shoot could be found. There is no later record from here despite several searches. A single seedling occurred in 1965 on the beach at Easter Quarff, South Mainland, at a point where it could not possibly survive. Sea pea is likely to be found again, somewhere, when a seaborne seed germinates and succeeds in forming a new colony.

L. pratensis. *Meadow Vetchling.* Frequently seen in rough grassy places (neglected roadsides, field- borders, etc.), rarely on holms in lochs or on low rocky sea-banks.

*Pisum sativum. *Garden Pea.* This is merely a very rare casual of rubbish-dumps (Lerwick, 1968), rubbly ground (Holmsgarth Ferry Terminal, Lerwick, 1985), and once in a sown grass field in Fetlar (1928).

*Melilotus indicus. *Small Melilot.* A very rare casual. Garden weed, 'Easterhoull', near Scalloway, 1976; one plant on the Lerwick rubbish-dump (by the Loch of Clickimin, 1980); garden of 'Lower Stonybreck', Fair Isle, a single plant (2006, P. Thomson).

*Medicago sativa subsp. **sativa**. *Lucerne.* A very rare casual twice recorded from South Mainland: cornfield near Ringasta, one plant (1962); sandy fields to the W of Exnaboe (1966).

*M. polymorpha. *Toothed Medick.* An even rarer casual: once recorded, as a garden weed, 'Easterhoull', near Scalloway (1976).

Trifolium repens. *White Clover.* A common plant of pastures and grassy places generally; also in much wetter ground such as saltmarshy turf, flushes, marshes, etc. Scott & Palmer (1987) write, 'White clover, in several agricultural strains, is a vital constituent of reseeding mixtures and is extensively employed in surface seeding and hill reclamation operations, as well as for hay and silage crops.' The writer finds it impossible to distinguish between the native and imported forms. However, white clover cannot be other than native in wet uninhabited and uncultivated places far from any obvious sign of past or present human activity.

*T. **hybridum** subsp. **hybridum**. *Alsike Clover*. Casual. Formerly an occasional relic or waif of cultivation which rarely survived in one place for more than a year or two. It was last noted in 1985: a patch on the roadside verge by the Dunrossness Primary School, South Mainland (J. S. Blackadder).

*T. **dubium**. *Lesser Trefoil*. Nowadays an uncommon casual. Apart from two areas in South Mainland where it has grown in sandy pastures, this is now a casual of reseeded ground: roadside embankments and verges; amenity sowings around new buildings; and (especially in South Mainland) in sandy arable ground. In 1924 it was seen on the links behind the Bay of Quendale, South Mainland. There then followed a long gap when no surveys took place. In the 1960s it appeared over a large area behind the bay, but by 1982 it had visibly declined, and by 1999 could not be found. In the 1950s and 1960s it grew in sandy pastures about Sumburgh Airport, also in South Mainland; by 1979 it was very scarce, and twenty years later could not be located, except as a rare casual in nearby arable land. The writer—bearing in mind that W. H. Beeby did not record it at all from Shetland, and the comment by Spence (1914) that in Orkney it was 'probably introduced, but found on natural pasture'— is encouraged to regard it at Quendale and Sumburgh as a former, long-persistent, but nevertheless temporary, well-established alien. Its decline at Sumburgh coincided with extensive works at the airport; however, no such disturbances have occurred at Quendale other than overgrazing.

T. **pratense**. *Red Clover*. A frequent plant of dry grasland, low sea-banks, roadsides, etc. It is also imported (as subsp. *sativum*) in seed mixtures for silage and hay.

***Thermopsis montana**. *False Lupin*. A well-established alien in one place in Fetlar: on the site of a derelict garden, 'Leagarth House', Houbie. It is not one of the relics of the original garden, but appeared *c*.1978 as a small patch which now (2008, WS) extends to some 150 squares metres.

***Lupinus × regalis** (*L. arboreus × L. polyphyllus*). *Russell Lupin*. Casual. A very popular garden plant which is occasionally seen outside cultivation where it has been planted; possibly sometimes as an outcast.

*L. **nootkatensis**. *Nootka Lupin*. Nowadays probably best regarded as a casual. A common garden plant which is sometimes seen outside enclosures as an outcast, straggler, or deliberate introduction. For a long time it was a feature of the banks of the stream between Tresta and Tresta Voe, Central Mainland, where it had become a well-established alien. However, in 2008 WS found that it had greatly declined and only a few clumps remained not far above the mouth of the burn.

***Laburnum anagyroides**. *Laburnum*. Casual, seldom seen outside enclosures.

***Cytisus scoparius** subsp. **scoparius**. *Broom*. Casual. Seldom, if ever, seen outside enclosures and probably not worthy of inclusion in this list.

***Ulex europaeus**. *Gorse*. Casual, capable of persisting for a long time. Occasionally seen by roadsides or near houses, a deliberate introduction in most, if not all, cases: roadside near lane to 'Gardie House', Bressay; lane to 'Wullver's Hool', Baltasound, Unst; by the N side of the B9076, opposite the entrance to Voxter Voe, N of Brae,

North Mainland; and in similar places elsewhere in Shetland. Around 1860 it was planted in a number of locations in the Veensgarth area of Tingwall, Central Mainland, and for a long time afterwards it was a feature of the area when in bloom. Eventually, from the 1960s onwards, it declined for various reasons; today (2008) only a handful of bushes remain.

*U. gallii. *Western Gorse*. Casual. One small bush on a rocky streambank at Vaadal, at the Fair Isle Water Pumping Station. A deliberate introduction.

Myriophyllum spicatum. *Spiked Water-milfoil*. Very rare and currently known from two lochs only, both in Unst. This was first found in Shetland in 1962 in the Loch of Hillwell, South Mainland, but could not be found some twenty years later. During this period the loch suffered badly from increasing eutrophication. In 1997 a Scottish Natural Heritage loch survey team located it in the Loch of Snarravoe, Unst, where it was seen by WS four years later. Also in 2001 a considerable quantity was found drifting ashore at the S end of the nearby Loch of Stourhoull (L. A. Inkster and WS).

M. alterniflorum. *Alternate Water-milfoil*. Common throughout the county in lochs, streams, etc., in mainly neutral to peaty waters. Although normally submerged it can occur on wet gravel as a terrestrial state rooting at the nodes.

*Epilobium hirsutum. *Great Willowherb*. A well-established alien in one place: in a sluggish watercourse (where it was first noticed in 1984) running for 200 m NE from the A970/A971 junction near Tingwall Airport, Central Mainland, before joining the burn running into Strand Loch. Otherwise it occasionally appears on waste ground, by ditches, etc., as a garden outcast or relic, and sometimes as a deliberate introduction.

E. montanum. *Broad-leaved Willowherb*. Accepted by the writer as native on the basis of its fine occurrence in 1991 in a steep, shady and damp gully by the coast below the Neap of Foraness, entrance to Dales Voe, North Mainland (L. A. Inkster and WS). It is on record from two steep-sided burns: the Burn of Sundibanks, near Scalloway (1887) and the Burn of Eelawater, Northmavine (1979). Although possibly native in these two places, its presence may be linked to the dumping of garden rubbish from nearby houses. Elsewhere it is a common or frequent weed of gardens, walls, waste ground, rubbly places, etc.

E. × aggregatum (*E. montanum* × *E. obscurum*). Only once recorded: Fort Charlotte, Lerwick (1966).

E. × mutabile (*E. montanum* × *E. roseum*). Another hybrid with just one record: near the ship repair yard by the Lerwick waterfront (1973).

*E. obscurum. *Short-fruited Willowherb*. A local, well-established alien of roadside ditches; gardens; rough, rubbly places; old walls, etc. Not recorded for Unst or Fetlar.

E. × schmidtianum (*E. obscurum* × *E. palustre*). Another rare willowherb hybrid, once found at 'Ominsgarth', Sandsound, Central Mainlan (1966).

*E. roseum. *Pale Willowherb*. This is best, perhaps, regarded as a very weakly established alien which has been around in Lerwick (its only station) since 1960 when it was found at Alexandra Wharf and at nearby Fort Charlotte. It has occurred (nearly always sparingly) in gardens, on old walls and rocks, near commercial buildings and car parks, and on rubbish-dumps and rubble. Because it grew in a town it was always liable to be removed through new developments, amenity improvements, and the like; its ability to persist as a perennial was thus sometimes short lived. Its latest sighting was in 2005, some twenty-five stems at the base of the SW external wall of Fort

Charlotte, at the end of a narrow close between the Garrison Theatre and a local council buiding to the N, in the company of *E. montanum* and *E. obscurum* (L. A. Inkster and WS). In 2009 WS counted some seventy-five flowering stems at this site.

*E. ciliatum. *American Willowherb*. A local casual of gardens and other planted areas, rough unused ground about commercial premises, ditches, etc. Most records are from Lerwick or nearby, with outliers in Fair Isle; Mid Walls, West Mainland; Scalloway; Voe, North Mainland; Whalsay, and Baltasound, Unst. First recorded from Shetland in 1979 from the Holmsgarth Ferry Terminal, Lerwick.

E. palustre. *Marsh Willowherb*. A frequent to common plant of ditches, burnsides, marshes, and watery places generally, becoming tall when growing among supporting vegetation; occasionally on old stonework.

*E. brunnescens. *New Zealand Willowherb*. An occasional well-established alien which has been regularly turning up in new places since 1989 when it was first noted between Voe and Laxo, Central Mainland. It grows on gravelly roadside verges and embankments and is very capable of spreading into and competing with the native vegetation; also found on other man-made habitats such as open gravelly areas by harbours and commercial premises, in quarries, at the bases of walls and (sometimes) on the walls themselves. Occurrences vary from tiny patches to extensive areas of a distinctive pale coppery-green colour. It is now widely distributed from the Ward of Scousburgh/Twarri Field area, South Mainland, to the Reawick area, West Mainland; Lerwick; the top of the Ward of Bressay; the Loch of Huxter Pumping Station, Whalsay; Yell (several places); and to the Belmont Ferry Terminal, Unst. It is so far unrecorded from most of the West Mainland, and from Northmavine and Fetlar.

Chamerion angustifolium. *Rosebay Willowherb*. This very rare native is currently known from only three localities: ravine of the Burn of Geosetter, South Mainland; heathery sea-banks, W side of Gluss Isle; and on a holm in Roer Water, both sites in Northmavine. Three further native stations were known in the recent past: among peat-haggs at the source of the Burn of Sundibanks, near Scalloway (first and last seen in 1955); on a piece of detached moorland in Black Water, North Nesting, Central Mainland (1960, but gone by 1978 through erosion); S side of Firths Voe, Delting, North Mainland, on a sea-cliff (1965, but not seen again and perhaps lost through erosion). In the distant past, Edmondston (1845) recorded it from cliffs on the W side of Ronas Hill, Northmavine, but no later record is known. He also, in 1837, noted it at Burrafirth, Unst, as did R. Tate in 1865, and W. R. Linton a year later. Finally, J. Sim in 1878 recorded it from the E side of Burra Firth, in a deep gully leading to Saxa Vord, Unst. Nothing more is known about it from Unst, an island where it may well have been native. It is commonly grown in gardens and is often seen outside enclosures as a well-established alien (not always near houses), and on disturbed rubbly ground in the vicinity of new buildings in the N part of Lerwick.

*Fuchsia magellanica. *Fuchsia*. Casual. A popular garden shrub not uncommonly seen as an outcast , straggler, or deliberate introduction. Rarely bird-introduced on old walls.

Cornus suecica. *Dwarf Cornel*. Very rare. This is known only from high moorland in Foula, and from two low-lying sites in Yell. A patch of moderate size near a house, 'Viewfield', Otterswick, Yell, has been known since at least 1915 and for many years recently has been afforded a degree of protection; in 2002 the patch (covering about eight square m) was in good condition, but in 2008 (WS) it seemed to be deteriorating. The other Yell site is a peaty bank nearly at sea-level at the Loch of Galtagarth, in the S of the island. Here, since at least 1925, a patch existed in good condition for some forty years; by the 1980s, however, a marked decline had taken place and by 2008 only seven very short barren stems were seen by WS about two or three feet from the edge of a slowly eroding peaty bank. Its future in Foula is very secure, the same cannot be said of Yell.

*Ilex aquifolium. *Holly*. Casual. Two tiny bushes close together in a ditch by the NE side of the road at Unst Airport, Baltasound, Unst (2005, P. V. Harvey and M. G. Pennington). They could not be found in 2008 (WS).

*Euphorbia helioscopia. *Sun Spurge*. Nowadays, this is still a frequent and well-established alien in light sandy arable ground at the S end of South Mainland, as at Hillwell, Scatness, Exnaboe, and Grutness. Formerly, it grew in good arable ground elsewhere in Shetland (as on the limestone bands of Central Mainland) but the marked decrease in the growing of oats, potatoes, etc., has led to the situation where there is only one relatively recent record from outside South Mainland: in the Sandness area of West Mainland (1995, BSBI field meeting).

*E. peplus. *Petty Spurge*. A rare casual weed of gardens, long-persistent where conditions remain suitable. There are only five localised records: Boddam, South Mainland, abundant (1890); Scalloway (1920); Bridge of Walls, West Mainland (1959); Garth, South Nesting (1966–1984), and Voe (1966–1976), both in Central Mainland.

*Linum usitatissimum. *Flax*. A rare casual of gardens or near houses, and mainly of bird-seed origin, with recent records from Lerwick, Scalloway, and Gutcher (Yell). Once found (*c*.1921) on the Baltasound foreshore, Unst. In Edmondston's day (1830s and 1840s) it was a rare field weed, 'introduced with foreign seed'.

L. catharticum. *Fairy Flax*. Frequent in dry or wet, short grassland (often stony or rocky), on grassy heaths and in sandy coastal turf, and in wet mossy hummocks in flushes. It is usually found on limestone, serpentinite, or granite.

Radiola linoides. *Allseed*. This is a tiny and easily overlooked annual plant of very short, damp or muddy, barish, heathy pastures often (but not always) near the coast in Fair Isle, Foula, South Havra, East Burra, West Burra, and Papa Stour, from all of which there are relatively recent records. In Mainland it was seen in the Loch of Brow/Skelberry area (1890, 1924); at Scatness (1960); and near Scalloway (1911). The following is the most recent record from Mainland: common in the disused quarry at Voe, Central Mainland (2002, P. V. Harvey), but not seen two years later.

Polygala vulgaris subsp. **vulgaris**. *Common Milkwort*. Frequent on the better soils (notably limestone or serpentinite) in dry grassy or rocky places such as streamsides, sea-banks, and dune-pastures.

P. serpyllifolia. *Heath Milkwort*. This milkwort, unlike the preceding, is common in damp acid heathy pastures; on moors and rocks; and among wet sphagnum.

*Aesculus hippocastanum. *Horse-chestnut.* Casual, rarely seen outside enclosures.

*Acer pseudoplatanus. *Sycamore*. Casual. A popular and frequent tree in gardens, occasionally planted in semi-wild places. Often surviving by old yards.

*Oxalis articulata. *Pink-sorrel*. Casual. A small clump among dumped rubble near the beach W of Ham, Bressay, probably a garden outcast (1999, WS).

O. **acetosella**. *Wood-sorrel*. A rare plant of the N half of Mainland. Damp mossy hollows under rocks, and on moist shady steep-sided streambanks, usually with ferns and mosses. It occurs at Djupa Gill, N of Voe; Burn of Valayre, near Brae; and in the ravine of the burn from Mill Loch, near Swining, all in North Mainland. In Northmavine it is recorded from the Beorgs of Collafirth (first and last seen in 1961); Beorgs of Housetter; Beorgs of Skelberry; and the Beorgs of Uyea (where it was very rare in 2000). Easily overlooked when scarce and not flowering, as is often the case. At Kergord Plantations, Weisdale, Central Mainland, it has been a well-established alien for a long time, especially in the Lindsay Lee wood where it is abundant.

*O. **incarnata**. *Pale Pink-sorrel*. Casual. Planted among rubble at the foot of New Street, Scalloway. It was first noticed in 1978, and survived until 1987 when it was obliterated by attempts to make the area look more attractive.

*Geranium endressii. *French Crane's-bill*. Casual of garden origin. Once found (1967–1973) as a garden relic on the site of a demolished house, Fox Lane, Lerwick, and (in 1982) as a garden outcast by Westerloch Drive, also in Lerwick.

*G. **pratense**. *Meadow Crane's-bill*. Casual, sometimes persisting for a long time. This very popular garden plant is often seen as a straggler or outcast near houses.

*G. **dissectum**. *Cut-leaved Crane's-bill*. Nowadays, this is probably best regarded as a casual. In the past, however, it seemed to be a well-established alien on two separate limestone outcrops in Scalloway before being eradicated from both by new buildings: near the castle (from at least 1956 to 1987); and near the school (from at least 1961 to 1989). There are six records of it as a casual: garden weed, Baltasound, Unst (1921); one plant within the enclosure of a reservoir near Scalloway (1961). The next three are all from South Mainland: one large plant by a farm building, Maywick (2002, J. S. Blackadder); garden weed, 'Sirius Cottage', Exnaboe (1999, T. Russell); four plants at the edges of cultivated fields, North Exnaboe (1999, 2000, TR). Several plants on agricultural ground, Brae, above Ham Voe, Foula (1999–2003, S. Gear).

*G. **ibericum**. *Caucasian Crane's-bill*. Casual. Garden outcast or escape, perhaps also a deliberate introduction. Dry grassy places by or near houses, roadsides, waste ground, etc. Perhaps surprisingly, all of its records come from South Mainland (Cunningsburgh and Sandwick districts), and one from Ireland (1996, R. C. Palmer).

*G. **× magnifcum** (*G. ibericum × G. platypetalum*). Status and habitat details are as for the preceding. Occasional in South, Central, and West Mainland, and in Unst.

*G. **molle**. *Dove's-foot Crane's-bill*. Probably best regarded as a rare but well-established alien in the past; today, however, apart from Grink Holm, between the island of Vementry and Mainland (where it was plentiful in 1992), it is little more than a casual. It used to be found in dune-pastures and in sandy arable fields in the Ringasta/Exnaboe/Sumburgh areas of South Mainland in the 1950s and 1960s, but now only as a rare casual. It also occurred in similar habitats in the N of both Yell and Unst, but apart from a 1997 record by R. C. Palmer from the Houlland area in the N of Yell, there are no recent records from these islands. Changing agricultural practices have led to its decline as a weed of arable ground. *G.molle* also favoured rocky outcrops, especially on limestone: Scalloway, two sites, with *G. dissectum* in each, both destroyed by building operations in the late 1980s; Brough, South Nesting, Central Mainland (last seen in 1961); rocky hillock to the NE of Seggi Bight, Vementry (last seen in 1984); by the N side of Papil Water, Fetlar (last seen in 1982);

and near the shore, Sand Wick, Unst (last seen in 1981). It occasionally appears as a casual in gardens and cultivated fields, etc.

G. **robertianum**. *Herb-Robert.* This is known from one site only: the shingle beach at Boddam, South Mainland, where for a very long time it grew abundantly by and to the N of the ruined house on the beach. In the mid-1990s a very unfortunate soil dumping and levelling operation substantially destroyed most of its habitat. Luckily, the plant began to make a partial recovery. In 2006 WS counted *c*.200 plants, many large and all in flower, over a narrow strip *c*.100 m long by the seaward edge of the site. However, in 2009, the strip had been replaced by a choking growth of *Atriplex prostrata* sensu lato (with some *A. littoralis* interspersed) and this had completely eliminated the herb-Robert except for two small patches, one at either end, and totalling some thirty-five plants. In 2009 WS noticed about fifteen occurrences on the unspoiled stony-bouldery beach directly opposite and on the other side of the *voe*, and in an open habitat entirely to its liking. It is to be hoped that it will thoroughly establish itself at this comparatively recent site because, at present, it is a mere shadow of itself at Boddam. The plant was first noticed at Boddam in 1890 by W. H. Beeby. It may, however, have been seen there much earlier, in 1769, by J. Robertson in whose manuscript list of unlocalised Shetland plants it appears. We know that he visited the S end of Mainland more than once during his tour. Herb-Robert has been seen as a casual, probably of garden origin, on a few occasions elsewhere in Shetland. Anon. (1949) states that herb-Robert 'has been found growing in Mid Yell'. Nothing more is known about this record.

*__G. phaeum__. *Dusky Crane's-bill.* Almost certainly a former casual of garden origin. There are two nineteenth-century records: one from Fetlar, the other unlocalised.

*__Erodium cicutarium__. *Common Stork's-bill.* Nowadays, a very rare casual. This is best regarded as having been a well-established alien in the latter half of the eighteenth century when G. Low, during his tour of 1774, recorded it as being plentiful in the sandy soil of Sandwick parish in South Mainland. This parish includes Levenwick and over a century later, in 1900, W. H. Beeby noted it from sandy ground near the sea at Levenwick, thus raising the possibility that Low had seen it here and not at nearby Sandwick where the habitat is less extensive. Since then there have been only casual records, each of one plant except where otherwise stated: near Exnaboe (1966), between the Loch of Quendale and Hillwell (1981), and S of Grutness (1993, P. V. Harvey), all three in sandy arable ground in South Mainland; cornfield, Uphouse, Bressay, frequent (1981); sandy beach, Rea Wick, West Mainland (1976); and in the yard of 'Braeview', Camb, Yell (1991, M. Spence).

*__Limnanthes douglasii__. *Meadow-foam.* Casual. A showy garden annual which is occasionally seen as an outcast near gardens, in old quarries, etc. It is capable of persisting and spreading, at least for a time.

*__Tropaeolum speciosum__. *Flame Nasturtium.* Casual. This very attractive climber was planted prior to 1950 among planted honeysuckle and other shrubs on steep rocks by the roadside near Vementry (Mainland). For many years, up to at least 1988, it grew and flowered well but by 1999 it was in rather poor condition and had stopped flowering. Not seen again.

*__Impatiens parviflora__. *Small Balsam.* Casual. Once recorded: garden weed, Garth, South Nesting, Central Mainland (1986).

*I. glandulifera. *Indian Balsam*. A casual of garden origin which is occasionally seen in the vicinity of houses, on dumped soil, rubbishy foreshores, etc., as an outcast, straggler, or deliberate introduction. A good site exists in the Souther House, area, Scousburgh, South Mainland.

*Hedera helix subsp. **helix**. *Ivy*. Casual. This is sometimes observed as a deliberate introduction on the walls of ruined houses. Around 1934 it was planted on a large rocky outcrop NW of Aithsness, West Mainland, where it merely survived for many years. In 2005 the single shoot was about one metre long; perhaps, at last, it was starting to expand.

Hydrocotyle vulgaris. *Marsh Pennywort*. A common species of watery or damp places: marshes, bogs, sides of burns and ditches, dune-slacks, etc.

[**Eryngium maritimum**. *Sea-holly*. Believed to be extinct. There are four records for this seashore plant, all from the nineteenth century and none supported by a later record: near Tangwick, Northmavine (prior to 1845); E shores of Bressay (prior to 1845); shingly ayre below Colvister, Yell (*c*.1855); and near Fitful Head, South Mainland (1884). In the case of the latest record the plant probably grew on the extensive sands of the Bay of Quendale.]

*Anthriscus sylvestris. *Cow Parsley*. This species is likely to be a well-established alien rather than a native. It is frequently found in or near the inhabited areas: grassy overgrown places, waste ground, about yards, by waysides, and in gardens, etc.

*Scandix pecten-veneris. *Shepherd's-needle*. Almost certainly a casual in the first instance, undoubtedly so in the second case: an unlocalised Shetland record devoid of further detail (probably 1890); a bird-seed casual in the garden of 9 Knab Road, Lerwick (1997, 1998, 2000, I. Clark).

*Myrrhis odorata. *Sweet Cicely*. An occasional casual of garden origin, sometimes seen by houses and burns, in old yards, etc., and capable of persisting for a long time. Most records are from the S of South Mainland, with a few scattered occurrences in Central Mainland and in Yell and Unst.

*Conopodium majus. *Pignut*. A well-established and widespread alien which often occurs in abundance in pastures, meadows, and relatively rich grassy places generally. It is particularly noticeable about Cunningsburgh, South Mainland, and Burravoe Yell. First recorded from Shetland, in 1865, from Voesgarth, Unst.

*Pimpinella sp. A species of *burnet-saxifrage*. Casual. Garden weed, Eastshore, near Sumburgh, South Mainland (1977). Its precise identity has been the subject of some debate.

*Aegopodium podagraria. *Ground-elder*. A well-established, frequent, unwelcome, and almost ineradicable alien weed of gardens, occurring as an outcast on waste rubbly ground, rubbish-dumps, heaps of dumped soil in old quarries, by burnsides, etc., and soon concealing large areas with its greenery.

Berula erecta. *Lesser Water-parsnip*. Native in one site, a well- established alien in another. It grows at both ends of the limestone valley which extends from the head of the East Voe of Scalloway N to the Strand Loch, all in Tingwall parish, Central Mainland. It was first recorded in 1888 from the burn from the Loch of Asta. In the mid 1950s, and until it was eradicated by a drainage scheme in the late 1970s, it grew from the bridge over this burn down to the head of the afore-mentioned *voe*. (Whether or not it ever occurred above the bridge is not known.) In 1986 material from the

Burn of Strand was introduced into the burn. (See later in this paragraph.) This move proved to be a pleasing conservation success. In 2008 the writer noted many patches above and below the bridge, one or two being up to ten m long. Clearly, lesser water-parsnip was flourishing as an alien where once it had occurred naturally. In 1966 it was recorded from the Strand Loch, but here, although likely to have appeared every year since, it never thrived until quite recently. In 2008 WS discovered several large but diffuse patches in sheltered embayments by the W and SW sides of this slightly brackish loch, usually among emergent vegetation, and in a locality which in 2000 supported a mere handful of tiny plants. It can only be assumed and hoped that a change in the water quality has encouraged *B. erecta* to establish itself more securely by the Strand Loch. It is always likely to be present there as a result of the fragments which wash down from its best station, the fine colony in the nearby Burn of Strand. This colony came to light in 1973 when the plant was found to be abundant and flourishing in the final 250 m stretch of the Burn of Strand before it enters the Strand Loch. In 1986 this site was also being threatened by a drainage scheme. Luckily, its plight had been noted and, before any work began, a number of conservation volunteers removed many clumps of the plant. Most of these were relocated very near to their original sites (sometimes alongside native stock), in areas not being affected by the digging. Despite the low number of clumps actually surviving the replanting—the remainder either dying or being washed away—the exercise was a success. In late 1986, after completion of the works, twenty-five clumps of native or relocated stock remained; by 1993 the figure had risen to fifty. In 2008 WS found many, often large, patches for a third of the way downstream from the sharp bend in the burn. The remainder of the the rescued plants were relocated elsewhere in the valley as follows: SE side of the Strand Loch; NW side of the Loch of Tingwall and adjacent roadside ditch; and in the burn from the Loch of Asta. Apart from the introduction at the last-named site these replantings failed after a few years.

Conium maculatum. *Hemlock*. Casual. First recorded in 1959 on waste ground near a timber merchant's premises, Freefield, Lerwick. A few plants occurred in the same general area in 1973, 1978, and 1984, after which the hemlock was not seen again. In 1968 it was observed on the Lerwick rubbish-dump (on the site of the North Loch), but no later record from here exists.

Bupleurum subovatum. *False Thorow-wax*. A very rarely recorded casual: in a yard, Symbister, Whalsay (1938); garden weed, Houl Road, Scalloway (1969); and on the Fair Isle Bird Observatory seed dump (1999, N. Riddiford).

Apium inundatum. *Lesser Marshwort*. Small shallow lochs and pools (some liable to dry up in summer), watery hollows and swamps, and in sluggish burns and ditches. Rare generally, occasionally locally frequent. Recorded from Fair Isle; Papa Stour and the nearby Holm of Melby; Esha Ness, Northmavine (especially in the W part of the peninsula); and Uyea, off Unst, plus the following two new sites: in the extensive swamp behind the storm beach WNW of the Loch of Quoy, Westing, Unst, abundant and probably the best site in Shetland (2002, WS); and by the shore of the more northerly of the two contiguous lochs NW of the Loch of Whitebrigs, near Burga Water, West Mainland, two or three plants (2006, N. Aspey).

Ammi majus. *Bullwort*. A very rare alien. A scarce garden weed at 7 Twageos Road, Lerwick, perhaps a bird-seed casual (1991); and one plant on a reseeded roadside embankment by the (old) A970 above Channerwick, South Mainland (1999, T. Russell).

*Carum carvi. *Caraway*. Formerly a well-established alien of dry pastures and banks (often near houses or churches or near the sea), and on sand-dunes. While still widespread it would appear to be slowly declining. The latest sightings are: near the beach SE of Little-ayre, Muckle Roe (1999, WS); and at West Sand Wick, Yell, abundant (1999, F. H. Perring).

Ligusticum scoticum. *Scots Lovage*. A widespread, frequent and sometimes locally common plant of sea-cliffs and rocky *geos*; much less often on beaches (as on the shingly beach at Symbister, Whalsay); rare on skerries (as on Bound Skerry in the Out Skerries); and on coastal stonework. A pleasing plant on account of its whitish flowers, glossy green leaves turning yellow in autumn, and its magenta stem bases.

Angelica sylvestris. *Wild Angelica*. A very familiar plant of damp or wet habitats: meadows; pastures; burnsides; holms in lochs, and by loch margins where it is often small and flowerless; also on the less exposed sea-cliffs and in rocky *geos*.

A. **archangelica**. *Garden Angelica*. The writer is inclined to regard this as a very rare and possibly overlooked native species, based on its apparently natural occurrences in three places on sandy or rocky shores. He feels that the likelihood of seeds drifting from Faeroe (where it is common and widespread on sea-cliffs) and occasionally germinating on our shores should be regarded as a possibility in the absence of any other explanation for these isolated occurrences. The three records are as follows: two massive flowering stems together (perhaps from two separate plants) and a flowerless plant some thirty m away, all among the sand-dunes, Nor Wick, Unst (1989, one plant in 1990 and one in 1991, no later record); one flowerless plant on the beach at the Sand of Meal, near Hamnavoe, West Burra (1993, 1994, 1995, but not seen again); and one flowerless plant among large rocks on the beach between Aith Ayre and Stebi Geo, Aith Wick, Cunningsburgh, South Mainland (2002, L. A. Inkster, not seen again). (In 1924 and 1954 it was recorded from a few townships in South Mainland. These records should be discounted as the plant was almost certainly cultivated at one time, and there is no evidence that they referred to plants growing outside gardens or yards.)

*Levisticum officinale. *Lovage*. A very rare casual which still persists by the ruins of Brough Lodge, Fetlar, where it has survived, probably as a kitchen garden relic, since the middle of the twentieth century or earlier. Otherwise a single plant in a garden near Scalloway Castle (1981).

*Pastinaca sativa var. **hortensis**. *Parsnip*. Casual. A mere garden outcast as by the roadside near Scalloway Castle (1975, R. C. Palmer).

*Peucedanum ostruthium. *Masterwort*. This scarce and weakly-established alien is an outcast or relic of cultivation about old yards, crofts, by burns near human habitation, etc. Long-persistent above the Burn of Sevdale, and at Williamsetter, both in South Mainland. In 1806 C. Fothergill listed it from Fair Isle, Papa Stour, Unst, and Fetlar.

*Heracleum sphondylium. *Hogweed*. Common. This is more likely to be a well-established alien than a native. Rough, overgrown, grassy and often rocky places (especially on the better soils such as limestone); dry pastures; among dunes; low sea-banks; and about churchyards and roadsides. Rarely far from human activity.

42

*H. mantegazzianum. *Giant Hogweed*. Formerly a long-persistent casual. From the mid 1940s to 1997 this garden outcast or escape grew by or near the roadside (occasionally on both sides of the road) by 'Leagarth House', Houbie, Fetlar. Once a feature of the local flora, it began to decline and by 1997 only one small plant remained. This was removed to a garden in an attempt to save and propagate the Fetlar stock; however, it died soon afterwards. In 1958 a few plants grew in a field near Houbie from where there is no later record.

*Daucus carota subsp. **sativus**. *Carrot*. Casual. Rarely seen as an outcast or relic of cultivation.

Gentianella campestris. *Field Gentian*. Still widely distributed throughout Shetland but not in the same numbers as in previous decades. Dry, turfy, herb-rich, often stony or rocky pastures, short coastal turf, dune-pastures, roadsides, etc. In 2005 it was pleasingly abundant by the ruins of 'Burg', W side of Muckle Roe (T. Harrison.)

G. amarella subsp. **septentrionalis**. *Autumn Gentian*. This is a native as well as an alien. As a **native** it is found locally in dry or slightly damp dune-pastures or (rarely) on banks bordering a sandy beach. It also grew, usually some way inland, in dry rocky pastures in a handful of places (mostly on limestone) but has not been seen in this habitat for many years until a new station was discovered, at Burravoe, Yell (see below). In recent decades it has suffered very badly from overgrazing, sand extraction, and agricultural and industrial developments generally. On the links below Scousburgh, South Mainland, it may be extinct, and behind the Bay of Quendale in the same parish, and at West Sand Wick, Yell, its numbers have been severely reduced. This fine plant, however, can still be found in four main areas: the southern end of South Mainland, and parts of Yell, Unst, and Fetlar. It is also on record from the two small islands of Balta and Huney, both off Unst. The following new records have recently come to light: Sandness, West Mainland (1953, M. A. E. Richards), no later record; HU3630/Q, almost certainly intending Banna Minn, West Burra (1956, N. M. Pritchard), no later record; Mousa, N side of West Pool, eight plants (2002, R. Norde). In 2009 L. Farrell found a fine colony of numerous plants in rocky herb-rich pasture near the house known as 'Skibhoul', Burravoe, Yell; although close to the sea it is not the usual sandy coastal habitat, and is perhaps now our only such site.

As a usually short-lived **casual** apparently introduced in sand extracted from the coast, autumn gentian has been noted as follows (recorders' names not listed here): by a track in a field (with wartime remains) N of the Burn of Hillwell, South Mainland, sparingly (1980–1982); five stations on stony or gravelly roadside verges of the A970 from NW of the Loch of Fladdabister to near the Lerwick Observatory, with as few as nine plants at one site to at least two hundred in another (1996, 1999, 2003, 2007, various recorders); sparingly by the roadside from the S side of Voxter Voe to the Sella Ness turn (2007–2009); one plant on a roadside verge, Graven (1991), no later record; thousands of plants on levelled gravelly ground close to the Sella Ness port complex, Sullom Voe, (1997), not one seen three years later. The last three sites are all in North Mainland. In Northmavine there is one record: hundreds of plants on gravelly ground within the perimeter fence of the Eela Water Water Treatment Works (1997), *c*.100 plants three years later. It has also appeared as a casual by the roadside above the S end of the Loch of Snarravoe, Unst, frequent (2009), and sparingly by the N side of the talc quarry at Quoys, W of Haroldswick, also in Unst (2008).

Hyoscyamus niger. *Henbane.* Casual. One plant among *Urtica urens* in a sandy field between Quendale and the Loch of Hillwell, South Mainland (1980).

Lycopersicon esculentum. *Tomato.* A surprisingly rare casual with only four records: Scalloway rubbish-dump (1963); Lerwick rubbish-dump, on the site of the North Loch (1963); five or six plants in flower on the sandy beach near the mouth of the Burn of Bouster, below Grimister, Yell (2003, WS); and seven plants in flower on the beach below the waiting-room at the Toft Ferry Terminal (2003, I. Clark).

Solanum nigrum subsp. nigrum. *Black Nightshade.* A very rare casual with just one record: garden weed at 'Brekka', East Voe, Scalloway (1958).

S. dulcamara. *Bittersweet.* Casual. One plant on grassy roadside verge, W side of the A970, near the entrance to the Scord Quarry, Scalloway (2004, B. Johnson), still there six years later. It produces both flowers and fruits.

S. tuberosum. *Potato.* Casual. Not surprisingly, this is a frequent relic or outcast of cultivation around crofts and farms, on rubbish-dumps, waste ground, and on low, rubbishy sea-banks in crofting areas.

Datura stramonium. *Thorn-apple.* A very rare casual. One plant in a field near the Ollaberry Primary School, Northmavine (2006, R. Anderson, per A. Williamson).

Convolvulus arvensis. *Field Bindweed.* A long-persistent casual in the garden of 'Gardie House', Bressay (1924), still there (2009). How long it had been there before 1924 is not known.

Calystegia sepium subsp. sepium. *Hedge Bindweed.* A weakly established alien which is occasionally seen as a garden outcast near houses, among rough herbage, and by overgrown burnsides and roadsides where it can persist for some time.

Menyanthes trifoliata. *Bogbean.* A frequent plant, especially in peaty places. Soft-edged bog-pools and small lochs; sheltered embayments of large lochs; deep, usually slow-flowing burns; and in marshy or swampy places.

Nymphoides peltata. *Fringed Water-lily.* Casual, planted. In 1806, C. Fothergill wrote, '... in a small lake nearly overgrown with aquatic plants, near Lund in Unst'. The site may have been the Loch of Vinstrick, below Underhoull.

Polemonium caeruleum. *Jacob's-ladder.* Merely a rare garden outcast on weedy ground, rubbish-dumps, etc.

Phacelia tanacetifolia. *Phacelia.* Casual. A very rare casual. Twice recorded: in a Scalloway garden (1960), and in a garden at Channerwick, South Mainland (2008, M. Perring).

Lithospermum arvense. *Field Gromwell.* Another very rare casual, once recorded in a new garden in Scalloway (1963).

Pulmonaria officinalis. *Lungwort.* Casual. A popular garden plant which is sometimes seen in the vicinity of gardens as an outcast or deliberate introduction.

Symphytum officinale. *Common Comfrey.* Casual. A cream-coloured form was found as a probable garden outcast near a house, Gardie, Mid Yell (1961).

S. × uplandicum (*S. officinale × asperum*). *Russian Comfrey.* A well-established alien of garden origin, frequently seen as an outcast in overgrown grassy places around houses and crofts, and on waste ground and rubbishy foreshores.

S. tuberosum. *Tuberous Comfrey.* Another well-established alien comfrey of garden origin, but much less frequently encountered than the preceding species. Grassy spots among or near houses; by roadsides and walls; in old quarries; and under

44

planted trees. In 2007 several patches (some quite large) grew in a semi-enclosed field to the N of Scalloway Castle. Many years earlier, in 1959, the plant was recorded as growing sparingly near the castle; it is not known, however, whether or not it has had a continuous existence in the area.

*Anchusa arvensis. *Bugloss*. A well-established alien of arable ground, especially cornfields on light sandy soils in South Mainland, and elsewhere in the county where sand is a noticeable feature. Bugloss can still be found, but not so commonly as in the past owing to the marked decline in the growing of oats and other crops.

Mertensia maritima. *Oysterplant*. A plant of shingly or sandy beaches, and one of our most attractive species. Unfortunately, since it was first recorded by J. Robertson in 1769, and by G. Low in 1774, it has disappeared from most of its fifty recorded sites. Easter Quarff, S of Lerwick, was one of its finest stations, bluish from a distance when the oysterplant was in full bloom. By 1983 a serious decline in its numbers began; today, there is probably not one plant on the beach. There are a few reasons for this general decline in Shetland: overgrazing, extraction of beach materials, and severe wave action in storms. With one exception it grows on beaches; the one departure from this is in Yell, at Catti Geo, N of West Yell, where it was found on the almost sheer side of a stack-like rock in the *geo* (1997, Scottish Natural Heritage). Currently, oysterplant can be observed in some quantity in four places only: Fair Isle (where it is a comparatively recent arrival) and at three stations in Northmavine (at Stenness, Ura Firth, and by the sea-loch S of Housetter). The occasional seedling can be found, sometimes where there is no previous record of oysterplant. These have the potential to initiate new colonies, but, as we have seen, there are obstacles in the way of strandline plants, not the least of these being the sea itself which can be both distributor and destroyer.

*Amsinckia sp. *Fiddleneck*. A very rare casual. Norwick, Unst: two plants among a luxuriant growth of *Matricaria discoidea* in a field where potatoes had been grown a year earlier (2000, M. G. Pennington). It has been tentatively named *A. lycopsoides* by E. J. Clement, an acknowledged expert on casuals.

*Plagiobothrys scouleri. *White Forget-me-not*. Another very rare casual. One large and two small plants in sown grass by or close to the main road, Unst Leisure Centre (1989). A contaminant of reseeding mixtures, this North American plant was first noticed in the wild in Britain in 1974, in Caithness.

*Myosotis scorpioides. *Water Forget-me-not*. A widespread, frequent and well-established alien of garden origin. Ditches, burns, marshy and swampy areas, etc., nearly always in nutrient-rich places close to houses and crofts, or by burns which run through inhabited areas. It is particularly well-established about the Lochs of Spiggie, Brow, and Hillwell, all in South Mainland. It was in the Hillwell and Brow areas that W. H. Beeby collected it in 1890; the fact that he neither collected nor recorded it from elsewhere may indicate that it was not at all widedspread in his day. In the early 1920s G. C. Druce added only a handful of new stations. It is the writer's opinion that this plant does rather better in the wild than in gardens which may be too dry for its liking.

M. secunda. *Creeping Forget-me-not*. A common plant of watery places: ditches, burns, bogs and marshes. Frequently associated with *Montia fontana* and *Stellaria uliginosa*. Its pale sky-blue flowers set it apart from all other native Shetland plants.

M. laxa. *Tufted Forget-me-not*. A frequent forget-me-not of watery habitats: ditches, burns, marshes, etc., widespread but rarely in significant quantity.

*M. arvensis. *Field Forget-me-not*. A widespread and well-established alien which favours a variety of habitats: waste rubbly ground; arable fields; near or among the ruins of old houses; dry, often rocky pastures; roadside banks; shingly seashores and among boulders on storm beaches; and as a weed in gardens.

M. discolor. *Changing Forget-me-not*. Probably native even though many of its sites are close to past or present human habitation or activity. One place where it seems to have a reasonable claim to indigenousness is in Fetlar: here and there on dry grassy slopes among the high sea-banks at the West Neap, on the W side of the island (2000, WS). *M. discolor* occurs commonly in dry rocky grassland; arable ground; on waste open ground; and (rarely) in damp pastures where it is tall and branched.

*Lappula squarrosa. *Bur Forget-me-not*. A very rarely recorded bird-seed casual. Garden of 9 Knab Road, Lerwick (1997).

*Stachys sylvatica. *Hedge Woundwort*. Merely a casual of garden origin. A large patch by a roadside ditch below 'Nordrafluir', Biggings, Papa Stour (2003, WS).

*S. × ambigua (*S. sylvatica* × *S. palustris*). *Hybrid Woundwort*. A well-established alien which seems to have been formerly grown in gardens, but which now frequently appears as an escape or outcast. Its habitats include cultivated ground, burnsides, roadsides, swampy places, foreshores, waste ground, and neglected corners about houses and crofts.

*S. palustris. *Marsh Woundwort*. It would appear that this is probably best classified as a well-established alien. Occasional throughout the county in or by arable ground; by roadsides and streams; and in damp or wet pastures near houses. In 1999 R. C. Palmer noted a number of large patches near Garth of Susetter, N of Voe, North Mainland, and WS observed similar occurrences above Scarfa Taing, Muckle Roe (2004), and by the NE margin of the Loch of Spiggie, South Mainland (2009).

*Lamium maculatum. *Spotted Dead-nettle*. Casual. A very rare garden outcast at Kergord, Weisdale, Central Mainland (1967) and at Gulberwick, near Lerwick (1991, R. C. Palmer).

*L. purpureum. *Red Dead-nettle*. This is a frequent and well-established alien of cultivated and waste ground, rubbish-dumps, etc.

*L. hybridum. *Cut-leaved Dead-nettle*. Nowadays best regarded as a casual. Formerly, and possibly from as early as 1865 when it was first noticed in Lerwick (its only Shetland station), this was a reasonably well-established alien in or near the town on rubbish-dumps, in gardens, and on weedy roadsides, etc. It was in good condition in 1981 in at least two places, but in 1996 it was recorded as just surviving in Hill Lane, and by 1997 it was not seen at all in Lerwick. In 1999 R. C. Palmer found about twenty plants in a raised weedy border by a path below a house, 'Nia Roo', in the upper NE part of Gulberwick, S of Lerwick. It was still there in 2000 (WS).

*L. confertum. *Northern Dead-nettle*. A frequent and well-established alien of cultivated and waste ground (especially on sandy soils), rubbish-dumps, etc; it is less often seen than *L. purpureum*, and is a little less widespread.

*L. amplexicaule. *Henbit Dead-nettle*. Nowadays best regarded as a casual. First recorded in 1955, this was in the 1950s and 1960s a weakly established alien in or near sandy arable ground in South Mainland: Jarlshof; Brecks, Scatness; about Exnaboe (in fair quantity among potatoes, 1966); Ringasta (common among oats, 1962); and at Mail, Cunningsburgh. There have been no records from this area since 1987 when a few plants were seen near the Pool of Virkie (WS). Elsewhere there are three casual records, one from Sandness, West Mainland (1967), Lerwick (1973), and from 'Vaila Hall', Vaila (1980).

*__Galeopsis speciosa__. _Large-flowered Hemp-nettle_. A very rare and attractive casual. There are only three records: Burravoe, Yell, a few plants among cabbages (1973); two plants in a rye-grass field, Cott, Weisdale, Central Mainland (1984); and at least four plants in the garden of 'Bixter House', West Mainland (1986), two plants (2006, BSBI field meeting).

*__G. tetrahit__ sensu stricto. _Common Hemp-nettle_. One of our more common and well-established aliens of arable and waste ground, rubbish-dumps, and recently excavated heaps of soil, etc. Although still widespread, it is not so often seen now following the great reduction in the cultivation of oats and potatoes in recent decades.

*__G. bifida__. _Bifid Hemp-nettle_. The remarks under the preceding species also apply here, and to which may be appended the following comments. _G. bifida_ clearly likes poorer soils than those enjoyed by _G. tetrahit_ sensu stricto, and is also less common; in fields where they occur together the present taxon is often the rarer of the two.

*__Glechoma hederacea__. _Ground-ivy_. Casual. Occasionally seen as a garden escape or outcast near houses and crofts (especially on old walls), and capable of persisting for many years. Most records are from the S of South Mainland where, as elsewhere in the county, it is apparently becoming scarcer. Two recent notices are worthy of mention: by the ruins of 'Hoide', Burrafirth, Unst (early 1940s or earlier, F. Sinclair), two large patches (2002, WS); and an extensive patch on an old wall, Clavel, South Mainland (2003, WS).

__Prunella vulgaris__. _Selfheal_. A common plant of meadows and pastures (including dune-pastures), dry heathy ground, fine coastal turf, gravelly roadside verges, etc.

__Thymus polytrichus__. _Wild Thyme_. Common in dry (often rocky or stony) grassland, dune-pastures, on herb-rich heaths, serpentinite and granite fellfield, sunny, craggy rocks, and gravelly roadside verges and embankments.

*__Mentha × gracilis__ (_M. arvensis_ × _M. spicata_). _Bushy Mint_. A well-established alien in Foula. First recorded in the 1940s from the lower end of the Ham Burn, just above the head of the _voe_, as an escape or outcast from nearby cultivation, this has now become thoroughly established. In 2004 it occupied some sixty square m, mostly in two large patches. The variegated form, sometimes seen in gardens, grew as a casual in 1967 on the Lerwick rubbish-dump (on the site of the North Loch).

__M. aquatica__. _Water Mint_. Our only native mint. Burns, ditches, and the stony or marshy margins of lochs, especially in sandy areas (as in the S of South Mainland where it is frequent and locally abundant—the Loch of Clevigarth, NNE of Exnaboe, being a good example), and in the limestone valleys of Central Mainland. Elsewhere it occurs in Fair Isle; around Walls, etc., West Mainland; and in the SE of Unst (Ramnageo, Scolla, Hannigarth, and Sand Wick areas).

*__M. × piperita__ (_M. aquatica_ × _M. spicata_). _Peppermint_. An outcast or escape from cultivation, this is a well-established alien of burns, ditches, marshy ground, and by roadsides and garden walls; sometimes forming large patches. Local, and not on record from Whalsay, Fetlar, and Unst. Hairy forms are occasionally recorded.

*__M__. __spicata__. _Spear Mint_. Another outcast or escape from cultivation, and a well-established alien by burns and lochs, in marshy ground, around old houses, and on waste ground, undercliffs, foreshores, etc. Local, and a little more widespread than the preceding. In 1898 it grew at the Burn of Skaw, Unst, where it still persists to this day, near the mouth of the stream. Only the glabrescent form is on record from Shetland.

*__M__. × __villosonervata__ (_M. spicata_ × _M. longifolia_). _Sharp-toothed Mint_. A very rare casual with only one record: in 1960 a patch grew on waste ground adjacent to the Lerwick rubbish-dump (by the Loch of Clickimin). A garden outcast.

*__M__. × __villosa__ (_M. spicata_ × _M. suaveolens_). _Apple-mint_. A well-established alien in one place, a casual in another. At least two patches on waste ground near a house, Basta, Yell, dating back to the 1930s or earlier (2004, WS). Sparingly by a stream between Cott and Sound, Weisdale, Central Mainland, a garden outcast or deliberate introduction (1961), not seen again. Represented by var. _alopecuroides_ in both sites.

*__M__. × __rotundifolia__ (_M. longifolia_ × _M. suaveolens_). _False Apple-mint_. This mint, almost certainly first noted from Shetland in 1924, and a relic or outcast of cultivation, is a well-established alien by grassy roadsides, ditches and burns, around old crofts and houses, and on waste ground and rubbish-dumps. Nearly all of its records are from South Mainland (especially about the Cunningsburgh, Sandwick, Hoswick, Channerwick, and Levenwick areas). There are a handful of records from elsewhere: Lerwick (1965); Firths Voe, North Mainland (1976, 1984); Ollaberry, Northmavine (1979); and Burravoe, Yell (1981, 1994).

__Hippuris vulgaris__. _Mare's-tail_. This plant occurs locally in Yell, Fetlar, and Unst, and in the S of South Mainland; abundantly in some sites. Deep sluggish burns, moorland pools and nutrient-rich lochs and their associated swampy margins, rarely on damp sandy loch shores. In Unst it grows in the Burn of Caldback and elsewhere; in Yell it flourishes at the Waters of Raga (HU4791) and Shinniwers Dale (HU4987), etc; and (in Fetlar) in the lower end of the Burn of Northdale, and in mires near Funzie, a recently discovered station (1991, M. Tickner). It was apparently found in 1897 in the burn from the Loch of Asta to the East Voe of Scalloway. No later sighting is known, and there is some doubt about the record itself; see Scott & Palmer (1987) for a fuller account. In 1977 a few plants were noticed by the Loch of Hillwell. Some twenty years later it had increased enormously and virtually extended around the entire loch. By the NW corner of the Loch of Spiggie a very few plants were seen in 1981; by 2001 there were signs of considerable expansion. About eight stems grew by the Loch of Brow, E of the island (1996, P. M. Hollingsworth and C. D. Preston); no later record. Mare's-tail must surely be a relatively recent and natural arrival in South Mainland. It is inconceivable that here it has escaped notice in the past, given the high level of interest which this part of Shetland has received from many naturalists over a long period.

__Callitriche hermaphroditica__. _Autumnal Water-starwort_. Local but widely spread over Shetland from the Loch of Spiggie (South Mainland) to the Loch of Cliff (Unst). A number of stations exist in West Mainland, but in North Mainland, Northmavine, Yell, and Fetlar it is very scarce. It is essentially a plant of the clearer, less peaty lochs; very rarely in deep burns. Often abundant, and evidently so when autumn gales bring it ashore in masses.

C. **stagnalis**. *Common Water-starwort.* A common plant of burns, ditches and pools, usually in shallow water or in muddy places liable to dry out in summer. It avoids strongly peaty areas.

C. **platycarpa**. *Various-leaved Water-starwort.* The habitats for this plant seem to be similar to those given under the preceding species, except that the current taxon may prefer deeper water. Examination of the pollen is the only reliable way to tell it apart from *C. stagnalis.* The writer accepts only the ten records from Shetland, nearly all pre-1987, which were based on material passed by J. P. Savidge. The distribution of this species is very poorly understood; it may eventually prove be frequent, but seems unlikely to be common in our area.

C. **hamulata**. *Intermediate Water-starwort.* Another common member of the genus, with habitats similar to those of *C. stagnalis*, but with a liking for deeper water.

Plantago coronopus. *Buck's-horn Plantain.* An abundant maritime species of dry coastal sward (especially in sandy areas); among rocks and on gravelly slopes and old coastal stonework. It is very rarely seen well inland, as on rocks between the Smalla Waters and the Loch of Hollorin, West Mainland (1967), and on the stony roadside verges of the A968, extending intermittently for two to three km N from the Camb junction, Yell (2001, WS).

P. **maritima**. *Sea Plantain.* Another abundant coastal plantain: sea-cliffs, sea-banks, rock-crevices, coastal turf, shingly beaches, and saltmarshes. It is also an important member of our two main fellfield habitats, the bare granite debris of Ronas Hill, Northmavine, and the bare serpentinite debris of Unst. Elsewhere inland it grows on rocky outcrops, margins of lochs, flushes, peaty bogs (rarely), and by roadsides.

*P. **major**. *Greater Plantain.* As subsp. **major** this is a very common and well-established alien of waste, rubbly or trodden ground about crofts, farms, houses, commercial premises, and by roadsides, etc. *Subsp. **intermedia** has been recorded from the shores of the Lochs of Tingwall and Clickimin (both in Central Mainland); on damp sand, with *Carex maritima*, on the site of a winter loch behind the Bay of Quendale (South Mainland); and S of Springfield, Fair Isle (probably on the sides of ditches). Much more study of this subspecies is needed before any reliable comments can be made regarding its habitat preferences and distribution.

P. **lanceolata**. *Ribwort Plantain.* Abundant on the better soils and in a wide variety of dry grassy places: sea-banks, coastal turf and sand-dunes; pastures; grassy places among rocky outcrops; cultivated land; and on waste ground, roadsides, and old walls.

Littorella uniflora. *Shoreweed.* A mat-forming plant which occurs abundantly on the stony, sandy or muddy bottom of lochs and pools, and on their gravelly margins; also in swampy places and seasonally flooded depressions.

*Fraxinus excelsior. *Ash.* Casual. Very rarely planted in unenclosed situations. It is not one of our native species as claimed by Raeburn (1888).

*Ligustrum ovalifolium. *Garden Privet.* Casual, planted. One record: a small bush by Freefield Road, Lerwick. First noticed in 1976, and still present and flourishing in 2009 (WS).

*Mimulus guttatus. *Monkeyflower.* A well-established alien of ditches, burns, loch margins and adjacent marshy ground, watery hollows, etc.; also, but rarely, in much drier places, as on waste rubbly ground and old mortared walls. Frequent to common in the inhabited areas, especially where traversed by streams or ditches.

*M. × robertsii (*M. guttatus* × *M. luteus*). *Hybrid Monkeyflower.* A well-established alien of the inhabited areas where it grows in ditches, burns, and marshy ground. Rare, and probably sometimes overlooked as some of its forms are very close in general appearance to *M. guttatus.*

*M. × burnetii (*M. guttatus* × *M. cupreus*). *Coppery Monkeyflower.* This is another well-established alien mimulus of the inhabited districts, and which frequents the same watery situations as those preferred by our other members of the genus. Local to occasional, sometimes abundant as near Northdale, Fetlar. The normal Shetland form has a coppery-orange corolla and a petaloid calyx; the form with a normal calyx is much scarcer but could still be found N of Haggrister, Northmavine (2009, WS). Two other forms are on record, both with a petaloid calyx. The first was found near Vestinore, Cunningsburgh, South Mainland, corolla bright yellow, last seen in 1992 but in cultivation in Scalloway. The second, with an orange-yellow corolla, occurs in three places in West Mainland: near the West Burrafirth ferry jetty; in abundance near West Houlland, SE of the Bridge of Walls (1997, L. A. Inkster); and by the house on the Snarra Ness peninsula (2009, LAI and WS).

*M. nummularius. *Blood-drop-emlets.* Since at least 1951 this has been a usually well-established alien in one place, in ditches about Noonsbrough, West Mainland. In some years it appears in very small quantity. However, its long association with Noonsbrough, and the fact that the present owner of the land is encouraging its continued existence, seem good reasons for listing it as a weakly established alien. This is the '*M. luteus*' of Stace (1997).

*M. × maculosus (*M. luteus* × *M. cupreus*). *Scottish Monkeyflower.* Casual. There are two records, both made by the 1995 BSBI field meeting in Shetland: lower end of the Burn of Mail, Cunningsburgh, South Mainland, below the bridge; and at the Burn of Dale, West Mainland, 'near the little bridge at HU196529'. It has not been recorded again from either station. (The misleading note by Scott & Palmer (1999) about the plant being naturalised at Cunningsburgh should be ignored.)

*'M. × caledonicus' (Silverside, ined.). (*M. guttatus* × (*M. luteus* × *M. variegatus*)). Casual. From 1952 (or earlier) this was a well-established alien in the S part of Cunningsburgh, South Mainland, especially in ditches by the lower end of the Burn of Mail, close to the main road. In 1964 it was flourishing, along with *M. × robertsii*; their success, however, led to their downfall. Some time after 1964 the ditches were cleared to facilitate drainage, an operation which all but wiped out the present taxon. Its latest recorded sightings are as follows: a few plants on shingle by the Burn of Mail, below the bridge (1981); ditch in meadow SE of Vestinore (1981), and where one plant grew in 1992 (L. A. Inkster and WS); a few plants in ditch by track just W of Clapphoull (1981). Stace (2010) records this as *M. x caledonicus* Silverside nom. nud.

*Cymbalaria muralis. *Ivy-leaved Toadflax.* Casual. A single plant grew in 1965 on the Lerwick rubbish-dump (on the site of the North Loch), probably from a garden.

*Kickxia spuria. *Round-leaved Fluellen.* A very rare casual, once recorded as a weed in the garden of a house, New Road, Scalloway (1959).

*Linaria vulgaris. *Common Toadflax.* A rare casual of garden origin except for the small patch which turned up in 1985 as a casual at the oil-rig service base, Green Head, Lerwick. It has been planted on the roadside bank at Collaster, Sandness, West Mainland, and is occasionally seen elsewhere as an obvious straggler from gardens.

L. purpurea. *Purple Toadflax.* A very rare casual. A single plant in rough grassy herbage by the W side of the B9074, at the turn to Blythoit, East Voe, Scalloway (2008, WS).

L. repens. *Pale Toadflax.* A very rare casual of garden origin. In 1993 one plant was found in a field at Daisy Park, Baltasound, Unst.

L. maroccana. *Annual Toadflax.* Another very rare casual of garden origin, as on the Scalloway rubbish-dump (1961).

Digitalis purpurea. *Foxglove.* Casual. A popular garden plant which is quite often seen as an escape or outcast near houses, on and by old walls, in old quarries, and on waste or disturbed ground, sides of roads and tracks, etc. In some cases it is likely to have been planted, sometimes well away from houses. It has been a long-persistent casual at Kergord Plantations, Weisdale, Central Mainland; in 2008 L. A. Inkster and WS noted about fifty plants in one place in the Lindsay Lee plantation and one plant in the Leegarth wood.

Veronica serpyllifolia subsp. **serpyllifolia**. *Thyme-leaved Speedwell.* A common plant of damp grassland, arable fields, gravelly roadside verges and similar man-made ground, around derelict buildings, on old stone walls, and a weed in gardens.

V. officinalis. *Heath Speedwell.* Another common plant. This species clearly prefers dry places: fine turfy banks; stony heathy pastures; rocky outcrops, ravines and scree; gravelly roadside verges and embankments and similar man-made ground, and on old walls. A glabrescent bright green form of this with narrow, obscurely toothed leaves was found in Northmavine by R. C. Palmer in 1993 on the gravelly margin of the road which ascends to the top of Collafirth Hill from the main road. Similar, and perhaps identical, forms have been noted from Shetland in the distant past.

V. chamaedrys. *Germander Speedwell.* A well-established alien in the Tingwall valley, Central Mainland, where intermittent and often quite large patches occur on the grassy banks and verges of the B9074 from below Utnabrake N to a little E of the Tingwall church. It is at its most common between Utnabrake and the N end of the Asta Golf Course. The vast majority of its appearances are by the W side of the road. It had become thoroughly established by 1954 when it was first observed in the valley. Elsewhere it occasionally occurs as a usually persistent casual in grassy places near houses, churchyards, cemeteries, etc., as near the old part of the Mid Yell cemetery (2004, A. D. D. Gear).

V. scutellata. *Marsh Speedwell.* A plant of watery habitats: marshes, burnsides, ditches, and loch margins. Frequent in Central and West Mainland, but very rare in South Mainland. In Northmavine there is one recorded station: near Hamar (1991, R. C. Palmer). It can be found in Bressay, Whalsay, Unst, and Yell, and in 1991 was detected in the N part of Fetlar (M. Tickner).

V. beccabunga. *Brooklime.* Another speedwell which prefers similar habitats to those given under the preceding species. It is very much a plant of the limestone valleys of Central Mainland (as in the Tingwall valley between Asta and Gott), and on the N part of the White Ness peninsula, also on limestone. Elsewhere in Shetland it is recorded from several places in the vicinity of 'Gardie House', Bressay (1991, R. C. Palmer), and at Eastwick, Ollaberry, Northmavine (last noted in 1979). Perhaps the best station in Shetland was discovered in 2004 by WS: abundant for 250 m in the ditch by the W side of the A968, above Tofts Voe, North Mainland.

V. anagallis-aquatica. *Blue Water-speedwell*. The third in our trio of speedwells of watery situations. This species inhabits slow-flowing ditches and burns, marshes and pools. It is found in two areas. The first is in Central Mainland where it occurs locally in limestone valleys, as in the Gott area and elsewhere, and on the geologically similar White Ness peninsula. The second area is South Mainland where it grows at Cunningsburgh, and also on the rich sandy soils of the S end of this division, notably behind the Bay of Quendale and at the Loch of Clevigarth, NNE of Exnaboe.

V. × lackschewitzii (*V. anagallis-aquatica × V. catenata*). Material collected by W. H. Beeby in 1888 from the burn from the S end of the Loch of Asta, near Scalloway, is believed by J. Burnett to be this hybrid. Beeby's material is in the Natural History Museum, in London. The second parent has no definite record from Shetland.

V. arvensis. *Wall Speedwell*. A local to frequent plant of dry grassy niches among rocks, on old mortared walls and among mossy turf on wall-tops, waste disturbed ground, and in arable land. It is nearly always in very small numbers.

*****V. agrestis**. *Green Field-speedwell*. Nowadays a rare casual. This was formerly a reasonably well-established alien of (chiefly) cornfields on sandy soils, but is now best regarded as a casual following the great reduction in the growing of oats. It also occurred as a garden weed. The three latest records are as follows: sandy track through a hayfield, Breakon, Yell (1997, RCP); vegetable garden, 'Fairhaven', Castle Street, Scalloway (1999, WS); and a garden weed at 'Voe House', Walls, West Mainland (2000, WS).

*****V. polita**. *Grey Field-speedwell*. A very rare casual of waste ground, gardens, etc. In Scalloway it was first noted in 1955, and intermittently since; in 1999 it was in some quantity in the vegetable garden at 'Fairhaven', Castle Street. Otherwise there are only three records: cornfields, Ollaberry, Northmavine (1893); near the post office, Baltasound, Unst (1962); and by 'Vaila Hall', Vaila (1980) .

*****V. persica**. *Common Field-speedwell*. A casual weed of gardens, rubbly soil, waste ground, etc., with a scattering of records throughout Shetland. Only four of these are post-1987, the latest being from the above-mentioned vegetable garden in Scalloway.

*****V. filiformis**. *Slender Speedwell*. A garden plant which has certainly become a well-established alien in at least one place: in and about the churchyard at Tingwall, Central Mainland, where (when it was first observed in 1956) had already become thoroughly established. There are numerous patches between the main gate and the church. Elsewhere it occurs in lawns, churchyards, cemeteries, and on roadside verges and ditch banks, etc. The majority of its records are from South and Central Mainland. Only one record exists for Northmavine, and there are none at all from Whalsay, Fetlar, Yell, or Unst. It is perhaps not seen as often now as in the past. Nevertheless, on the verge and in adjoining grassland by the N side of Ingaville Road, Scalloway, where it was first noticed in 2005 (WS), it has been gradually increasing.

*****V. hederifolia** subsp. **hederifolia**. *Ivy-leaved Speedwell*. This is a formerly well-established alien, and although perhaps not yet a casual, is much scarcer now owing to loss of habitat, notably cornfields and other crops on sandy or limestone soils. Other habitats are waste open ground and gardens. It was frequent in South and Central Mainland, but rare or unrecorded elsewhere.

*****Hebe × franciscana** (*H. elliptica × H. speciosa*). *Hedge Veronica*. Casual. Almost certainly originating from seeds from nearby gardens. A fairly popular garden shrub.

One plant on rocks below Fort Charlotte, Lerwick, near the shops (1991), rising to at least ten plants (1999). Still present in 2005, with more on both sides of the S wall of the fort. In 1991 it also grew on the wall of a ruined house, Crooked Lane, Lerwick.

Melampyrum pratense subsp. **pratense**. *Common Cow-wheat*. This appears to be a very rare and elusive plant of steep N-facing upland heaths or moors (on present knowledge). First recorded, by G. C. Druce in 1921, a single plant high up on the NE side of Saxa Vord, Unst. It has never been noted here again. In 1971 A. Williamson came across it in the Black Butten area on the N side of Ronas Hill, Northmavine, about a score of plants, and in 1989 he saw three in the same area. On 17 July 1991 he, along with L. A. Inkster and WS , counted 110 specimens in flower, spread over three small areas centred on HU303841. Eight years later LAI, after a long search, found just one in flower.

Euphrasia. *Eyebrights*. A new in-depth study of the genus in Shetland is very much overdue. All *Euphrasia* species and some subspecies have been given English names in Stace (2010).

E. arctica subsp. **arctica**. Widespread in damp meadows, marshes, dry or damp grassy pastures, and on dry turfy roadsides, etc.

E. arctica subsp. **arctica** × **E. nemorosa**. Norwick, Unst (1921); Lund area, Unst (1995, BSBI field meeting).

E. arctica subsp. **arctica** × **E. confusa**. Probably frequent, and in drier grassland than that favoured by the first-named parent.

E. arctica subsp. **arctica** × **E. ostenfeldii** sensu lato. Hametoun, Foula (1963), and Ham, Foula (1965).

E. arctica subsp. **arctica** × **E. micrantha**. Green Burn, Sand of Sand, near Brough Lodge, Fetlar (1929); Out Skerries (1995, BSBI field meeting).

E. arctica subsp. **arctica** × **E. scottica**. Perhaps frequent, for both parents are not at all uncommon and share the same general habitats. East side of West Burra (1959); above 'Loot', Firths Voe, North Mainland (1965); by the E side of the East Voe of Scalloway (1966); and Dale Water, N of Clumlie, South Mainland (1978).

E. nemorosa. Frequent in fine dry, sometimes rocky grassland, dune-pastures, by roadsides, etc.

E. nemorosa × **E. micrantha**. Wick of Hagdale, Unst (1963); Grisigarth, Foula (1973, also 1982 (R. C. Palmer)); Keen of Hamar, Unst (1995, BSBI field meeting).

E. nemorosa × **E. ostenfeldii** sensu lato. West Voe, Housay, Out Skerries (1978).

E. nemorosa × **E heslop-harrisonii**. Given for Shetland by Stace (2010).

E. confusa. A common eyebright of dry, closely grazed grassland (especially on steep slopes and among rocks), and near the sea on cliff-tops and in dune-pastures.

E. confusa × **E. foulaensis**. Short turf, Scat Ness (1960), and cliff-top, Noss, SW of Spiggie (1960). Both locations in South Mainland. (The Collafirth Ness record given by Scott & Palmer (1987) should have appeared under *E. foulaensis* × *E. micrantha*.)

E. confusa × **E. micrantha**. Hagdale (1954), and stony pasture, Belmont (1960). Both stations in Unst.

E. confusa × **E. scottica**. Heathery grassland above Isbister, Northmavine (1959).

E. frigida. Known only from a few places on the high ground in Foula where it is associated with *Luzula sylvatica*, *Eriophorum angustifolium*, and *Anthoxanthum odoratum*. It should be looked for in similar habitats elsewhere in Shetland.

E. foulaensis. Widespread, especially in coastal turf with *Plantago maritima* and *Armeria maritima*; also (but less often) inland on open damp ground.

E. foulaensis × **E. ostenfeldii** sensu lato. As both species prefer coastal habitats, the hybrid is likely to be more widespread than the following four confirmed records would suggest: Isle of Fethaland, Northmavine (1959); Foula: near Ristie, and at the S end (1973); and on cliff-tops SW of Westerwick, West Mainland (1978).

E.foulaensis × **E. micrantha**. This hybrid is on record from a handful of sites, mostly from Fetlar and Unst, but with one record from Mainland and one from the North Isle of Gletness, off South Nesting, Central Mainland.

E. foulaensis × **E. scottica**. Ronas Hill, Northmavine (1952); Klifts, Wick of Tresta, Fetlar (1967); Toolie, near Tooa Stack, W side of Herma Ness, Unst (1995, BSBI field meeting). The Fetlar record is, however, somewhat doubtful.

E. ostenfeldii sensu lato. An occasional eyebright of short coastal turf (especially on cliff-tops), as well as inland on the Unst and Fetlar serpentinite. Although it is fairly widedspread—from Fair Isle to Unst and from Foula to Out Skerries—there are few stations on the E coast of Mainland. In Shetland this aggregate consists of two very similar species, **E. ostenfeldii** and **E. marshallii**. They are often represented by small condensed forms in highly exposed coastal sites; this makes it very difficult, if not sometimes impossible, to tell them apart. It is likely that *E. ostenfeldii* is the more frequent of the two.

E. ostenfeldii sensu lato × **E. micrantha**. Baltasound, Unst (1953); Grisigarth, Foula (1973); Out Skerries (1978); ravine of the burn from Helga Water, near Hillswick, Northmavine (probably 1979).

E. ostenfeldii sensu lato × **E. scottica**. Although no material has been unequivocally named as this, plants with shortly hairy leaves and flowers of medium size are very probably of this parentage. They have been observed by the coast and inland in the following places: grassy pasture, Colvister, Yell (1961); grassy ground by the shore below the post office, and in saltmarshy turf near the mouth of the Burn of Gerdie, both in Baltasound, Unst (both 1962); grassy slopes by a stream, Nissetter, near Ollaberry (1968), and among small rocky outcrops, The Dale, Urafirth (1979), both stations in Northmavine.

E. micrantha. This is a common and widespread eyebright of dry, or damp, heathy or moory places (frequent among heather), including gravelly tracks and roadside embankments in moorland areas; also on bare serpentinite fellfield and associated patches of mossy-heathy ground.

E. micrantha × **E. scottica**. Three records, all by the 1995 BSBI field meeting: Ronas Hill, Northmavine; by a burn between Mill Field and Mouslee Hill, Herma Ness; and in the Lund area, both in Unst.

E. scottica. A frequent eyebright of damp or wet habitats: by the marshy sides of lochs and burns; in rough damp pastures; and in flushes and boggy places.

E. heslop-harrisonii. Apparently very rare. First discovered, in 1969 by WS, on the Holms of Uyea-sound, off the NE corner of the island of Vementry. In 1973 K. G. Messenger found it in three places in Foula: rich pasture, Stoel, at the S end of the island; in hayfields and by roadsides, Ham; and in infield grassland, Loch. In 1996 D. Rae made a special study of this species in Shetland. On the Holms of Uyea-sound he found around 1,000 flowering/seeding plants, their identity confirmed by A. J.

Silverside, the *Euphrasia* expert. In Foula he could not find it at Stoel or Loch, while material from Ham was not confirmed by AJS as this species. Its current status in Foula needs further investigation, preferably after a new examination of the original 1973 material.

Odontites vernus subsp. **litoralis**. *Red Bartsia*. An attractive, rare and decreasing species of coastal (often sandy) pastures, and (rarely) arable ground near the coast. It is recorded from five areas: Fair Isle; about Sumburgh and near Okraquoy, both in South Mainland; near Hamnavoe, West Burra; and Sand Wick, Unst. It occurred in several places in Fair Isle, but was last seen, in 1963, by a cornfield near the Haa and in a hayfield at Taft (near Leogh). In the Sumburgh area it has been noted in several places. Virkie, at the head of the Pool of Virkie (1962); Eastshore (1960s). Since at least 1969 red bartsia has occurred in rich pastures SE of the Sumburgh Hotel, close to the S-shaped pond. It is important that this site continues to support this rare plant. In 2003 P. V. Harvey found about 400 plants close to the small pool near the jetty at Grutness Voe, a previously overlooked station. The following year he saw many hundreds of plants in the Wils Ness area of the airport; unfortunately, by 2005, most of these had vanished under airport expansion. In 1983 many plants were found in coastal turf at the head of the beach below Okraquoy, Cunningsburgh (first seen in 1962), but in 2000 only a dozen or so were seen (R. Riddington). In West Burra in 1959 it appeared on weedy ground near Setter, near Hamnavoe, but was not noted here again. In 1983 it grew commonly on rough thistly ground between Meal and the Sand of Meal, also near Hamnavoe, but only eight plants were seen in 1993 (L. A. Inkster). Finally, at Sand Wick, Unst, it was seen in three places behind the beach (2000, M. G. Pennington).

***Parentucellia viscosa**. *Yellow Bartsia*. A very rare casual: one plant among rough grasses and docks very close to the marina at Aith, West Mainland (1998, WS).

***Rhinanthus angustifolius**. *Greater Yellow-rattle*. Possibly a weakly established alien in T. Edmondston's time, but probably only a casual in W. H. Beeby's day and later. Edmondston (1845) recorded it from 'Yell and Northmavin, on peaty soil' and commented on how much it differed from *R.minor*. In 1890 WHB collected it in a cornfield near Ringasta, Dunrossness, 'apparently established'. It was last recorded by G. C. Druce, in 1920 and 1921, at Baltasound, Unst. One of his three sheets at Oxford is further localised to 'Buness'.

R. minor. *Yellow-rattle*. A common to frequent plant of grassy places. Subsp. **minor** and subsp. **stenophyllus** occur in meadows, cultivated land, marshes, on holms in lochs and by roadsides. Subsp. *stenophyllus* is likely to be the usual form of this variable species in lowland districts. Both it and subsp. *minor* have bright yellow corollas. Subsp. **monticola** prefers dry heathy pastures, especially on serpentinite, while subsp. **borealis** (which appears to be quite rare) frequents dry stony broken ground. Both of these subspecies have brownish-yellow corollas. These four forms are often difficult to separate; indeed, Stace (1997) suggests that recognition of the subspecies of *R. minor* 'may be better abandoned'.

Pedicularis palustris. *Marsh Lousewort*. Locally frequent in nutrient-rich marshes and swamps, and clearly preferring the less acidic soils favoured by the following.

P. sylvatica subsp. **sylvatica**. *Lousewort*. Common in wet or damp heathery pastures (often with *Molinia caerulea*), boggy ground, and on damp heaths.

Pinguicula vulgaris. *Common Butterwort*. A common and widespread species in open spots on wet heaths and moors; also by steep-sided burns and low sea-banks where water is oozing or dripping from the rocks.

Utricularia vulgaris sensu lato. *Greater Bladderwort*. A plant of usually deep pools (especially in peaty swampy areas), sluggish burns, and quiet corners of lochs. Frequent in West and Central Mainland, but much scarcer in other areas. The two segregates (*U. vulgaris* sensu stricto and *U. australis*) are best told apart on flowering material. However, this has been recorded once only (in the Loch of Beith, North Roe, Northmavine, in 1959 by C. J. Cadbury) who thought that his single flowering specimen was *U. vulgaris* sensu stricto. On the other hand, Druce (1922) was of the opinion that the Shetland plant was probably *U. australis*. Whether or not both grow in Shetland remains to be settled, and this uncertainty is likely to continue until more flowering material comes to light. Only then can the arrangement of the glands inside the spur be ascertained, the only reliable method of separation.

U. intermedia sensu lato. *Intermediate Bladderwort*. This element of the genus, rarer than the preceding, grows in swamps, mossy peaty pools and lochans mainly in West Mainland and about the Lochs of Spiggie and Brow in South Mainland. In other parts of Mainland it is rarely noted, and is unrecorded from Yell, Unst, and Fetlar. The aggregate consists of three segregates, *U. intermedia* sensu stricto, *U. stygia*, and *U. ochroleuca*. After painstaking examination of the quadrifid hairs on fresh material from a number of sites, L. A. Inkster has shown that ***U. stygia*** (Nordic bladderwort) is a member of the local flora.

U. minor. *Lesser Bladderwort*. Shallow peaty lochans, swamps, moorland pools and ditches. Frequent in West Mainland, thinning out elsewhere, and unrecorded or absent from South Mainland and Fetlar. Flowering was first reported in 1960 in the small loch by the road above Newton on the way to West Burrafirth, West Mainland. In 1995 WS found it in bloom in Wife's Water (HU3047), N of Easter Skeld, West Mainland, and in the same year L. A. Inkster saw it in flower in several West Mainland sites including the afore-mentioned loch above Newton. In Muckle Roe, in 2003, he recorded it flowering plentifully in a lochan just N of Muckla Water.

Campanula rotundifolia. *Harebell*. An apparently extremely rare native which may be on the verge of extinction, and one which is very difficult to detect when not in flower. There are only six genuine or acceptable records, as follows: moor between Skelberry and Boddam, South Mainland, a large patch in one place (1900), no later record; near Laxfirth, Tingwall, Central Mainland (before 1845), no later record; a flowering patch near Uphouse, Bressay (*c.*1930), no later record ; dry rocky heathy turf, Ness of Islesburgh, Northmavine, in one place (1946), last seen in 1991; near Helliers Water, Unst, three or four plants (*c.*1960), no later record; and in dry rocky turf, Skeo Taing, Baltasound, Unst (*c.*1930), last seen in 2000, a few tiny patches.

Jasione montana. *Sheep's-bit*. A common and often abundant plant of dry places: acid, often rocky, stony or heathy grassland; grassy sea-cliffs and eroding coastal banks; rocky streamsides; and on the sides of old quarries. Often accompanied by *Solidago virgaurea*, another lover of dry ground.

Lobelia dortmanna. *Water Lobelia*. Occasional to frequent throughout the county on the stony bottom of lochs and pools with relatively clear water. It avoids strongly peaty or soft-bottomed lochs, and flowers only in shallow water.

*Sherardia arvensis. *Field Madder*. Casual. Very rare. The following records are from South Mainland: roadside near Ringasta, very scarce (1957); oatfield by Loch of Hillwell (1965); garden weed, Virkie (1970); arable ground by the N side of the Pool of Virkie, and in the garden of nearby 'Sirius Cottage' (1995–1997, T. Russell). Elsewhere in Shetland it has been seen in a hayfield N of Tingwall churchyard, Central Mainland (1964); as a garden weed, 'Bakkakot', Port Arthur, Scalloway, two plants (1999, L. A. Inkster); and one plant in a hayfield, Houbie, Fetlar (1927).

*Galium odoratum. *Woodruff*. Casual. There is one reliable record: 'by rivulet at Baltasound [Unst]: probably a garden escape' (1903 or earlier).

G. palustre subsp. **palustre**. *Common Marsh-bedstraw*. Common and widespread in very wet places, namely, marshes, sides of burns, ditches, dune-slacks, etc., often straggling among taller vegetation.

G. verum. *Lady's Bedstraw*. This is a species of dry rocky grassland, especially on limestone or sandstone; steep rocky burnsides, grassy sea-banks, and sandy coastal pastures. It is most frequent in South and Central Mainland, occasional elsewhere, rare in Yell (except in sandy areas), very scarce in Unst and Fetlar, and unrecorded from Whalsay.

*G. mollugo subsp. **erectum**. *Hedge Bedstraw*. A very rare casual with only three relatively old records. Enclosure at Helliers Water, Unst, several plants (1962); small patch by edge of arable ground at the N end of the Loch of Spiggie (1963), and by the grassy margin of a hayfield, Easter Quarff (1966), both sites in South Mainland.

G.saxatile. *Heath Bedstraw*. An abundant plant of poor dry acid soils: heaths and moors, stony peaty places, rocky outcrops, and on gravelly roadside verges and embankments. In Orkney *G. sterneri* (limestone bedstraw) occurs in a number of places and should be searched for in suitable habitats in Shetland.

G. aparine. *Cleavers*. Frequent on shingle beaches throughout Shetland. Also, but rarely, as a weed in gardens and on waste barish ground near the coast.

*Sambucus nigra. *Elder*. Casual. This shrub or small tree is widely grown in gardens, and is frequently seen around crofts and houses and by burns. It is very hardy and persistent, and may often survive by or near a ruined crofthouse long after the roof has collapsed. In 1982 two bushes grew on the large holm near the E side of Punds Water, E of Hamar, Northmavine, obviously planted; however, in 2001 they could not be found.

*Symphoricarpos albus. *Snowberry*. Casual. A garden shrub which is occasionally planted near houses, also a probable garden outcast.

Lonicera periclymenum. *Honeysuckle*. A shrub of relatively inaccessible locations: steep-sided burns and ravines, crags, sea-banks and sheltered sea-cliffs, and holms in lochs. Often trailing among heather or over rocks. Local, but frequent in West and North Mainland. Occurs from the Burn of Geosetter (South Mainland) to the cliffs of Burra Firth (Unst). A popular garden shrub, and one which has been brought into cultivation from local native sites on many occasions. Consequently, the status of an occurrence on a rock or by a burn near houses would be difficult to ascertain without local knowledge.

*Valerianella locusta. *Common Cornsalad*. Probably best regarded as a casual even in the 1960s when it was last seen in its only two recorded and widely separated areas. In 1865 it was noted from sandy banks and fields, Norwick, Unst. It was still there in

1921, and in 1962 when it grew in small quantity among potatoes near the sea; no later record. In 1966 it was found in two places in South Mainland: by a hayfield between Scat Ness and Sumburgh Airport, very rare; and two plants in sandy fields at the W side of Exnaboe. No subsequent record is available from either place. It would be reasonable to assume that at one time it had been a fairly well-established alien at Norwick.

*Dipsacus fullonum. *Wild Teasel*. A very rare casual. Leaves (probably belonging here and not to *D. sativus*) were found on a Lerwick rubbish-dump in 1967, and in 1978 a single plant in flower was recorded from a garden in Bell's Road, Lerwick.

Succisa pratensis. *Devil's-bit Scabious*. A common species throughout Shetland in damp, often heathy pastures, meadows (damp or marshy), holms in lochs, low sea-banks, and on steep ungrazed rock-ledges. It varies greatly in size from dwarfed examples in short coastal turf to tall and luxuriant plants on well-vegetated ungrazed islands in lochs.

Arctium nemorosum. *Wood Burdock*. A very rare plant which has only one native station in Shetland, at Sumburgh, South Mainland; unfortunately, it urgently needs a serious long-term conservation programme if extinction is to be avoided. It grew, always sparingly, among the sand-dunes from the West Voe of Sumburgh to Grutness Voe. In 1978 an extension to the S end of the airport's main runway brought many long-buried seeds to the surface. These germinated and for a year or so the plant was abundant in one place, only to vanish when the ground reverted to closed pasture. Since 2000 the plant has become increasingly confined to the immediate area of 'Betty Mouat's Cottage' (near the Old Scatness Broch), notably on rubble and by the edge of an arable field. In 1967 this occurred as a casual in a garden at Quoyness, above the Loch of Strom, Central Mainland.

Saussurea alpina. *Alpine Saw-wort*. A very rare arctic-alpine which is now known from three widely separated localities. It was first observed in 1837 on Ronas Hill, Northmavine, where it grew very sparingly and elusively on the granite fellfield of the summits and slopes of Ronas Hill and neighbouring Mid Field. Its future seemed to be in doubt. However, a fine discovery was made by the 2006 BSBI field meeting: fifty-five plants on fellfield over about one square hectare SW of the source of the Burn of Black Butten, in HU3084, a square which also contains other rare plants such as *Empetrum nigrum* subsp. *hermaphroditum* and *Melampyrum pratense.* A new site, far from Ronas Hill, was found in 1962 on the NE slope of the Hill of Colvadale, Unst. In 2003 it was rediscovered by C. Geddes and A. Payne. Later that year L. A. Inkster and WS counted several hundreds of plants in short heathy pasture and associated stony ground. The site lies in the NE corner of HP6105 at *c*.105 m in altitude. The third station came to light in 2005 when C. Robson found two closely separated colonies, each of fifty to sixty plants, on flushed stony ground at *c*.200 m in altitude, above the SE corner of Smerla Water, SE of Gonfirth, Central Mainland. Apart from three records of flowering (1889, W. H. Beeby; 1958, J. Copland; *c*.1962, A. Williamson), all from Ronas Hill, only barren rosettes have been found. The writer always hopes that he will come across a flowering specimen one day.

*Carduus nutans. *Musk Thistle*. Casual. A single specimen was found, possibly in 1890, by A. H. Evans on a Shetland beach, 'very likely from the rubbish of boats'. According to Druce (1922) it grew at Balta Sound, Unst.

Cirsium vulgare. *Spear Thistle*. A common thistle of dry places: pastures, upper parts of beaches and adjacent grassy banks, gravelly roadside embankments and verges, rubbly waste ground, etc. Often misnamed 'Scotch thistle' in Shetland, a species which does not occur here except, possibly, in gardens.

*C. heterophyllum. *Melancholy Thistle*. Casual. This is occasionally cultivated in gardens and is sometimes seen in inhabited areas either as an outcast or a deliberate introduction. It has persisted near St Colman's Episcopal Church, Burravoe, Yell, for a very long time.

C. palustre. *Marsh Thistle*. A common thistle, often associated with *Juncus effusus* in wet, marshy and tussocky meadows and pastures, by moorland burns, in wet grassy places among the moors, and on coastal banks and roadside verges. It overwinters as a rosette which is often eaten by sheep at that time, despite the prickles.

*C. arvense. *Creeping Thistle*. A well-established alien if not a native. Common in rough grassy places, cultivated ground, and sandy pastures (especially in parts of South Mainland where it overruns large areas); also by field-borders, roadsides, old buildings, and on waste rubbly ground and rubbish-dumps. Several varieties occur in Shetland, including a soft-leaved, scarcely prickly form of sandy ground about the 'Haa of Houlland' and Breakon areas in the NE corner of Yell where it is common. The same form also grows in the sandy Exnaboe and Sumburgh areas of South Mainland.

*Centaurea montana. *Perennial Cornflower.* Casual. A popular species which is much cultivated in gardens, nearly always as the lilac-flowered form. It is sometimes seen on roadside banks near houses as an outcast or deliberate introduction.

*C. cyanus. *Cornflower*. A very rare casual. Cornfields, Shetland (early 1840s or earlier); cornfield, Levenwick, South Mainland, a few plants (1900); cornfields at Colvister, Yell (*c*.1935); three plants in a clover-field off Houl Road, Scalloway (1961); and two plants on a recently reseeded roadside verge by the E side of the East Voe of Scalloway (1991, B. Johnson).

*C. solstitialis. *Yellow Star-thistle*. A very rare casual with just one record: one plant among night-scented stock in the garden at 'Seaview', Okraquoy, Cunningsburgh, South Mainland (1980).

*C. melitensis. *Maltese Star-thistle*. Another very rare casual. In 1956 this grew on the Lerwick rubbish-dump (by the Loch of Clickimin).

*C. diluta. *Lesser Star-thistle*. The third member of our trio of very rare casual star-thistles, a single plant in 1961 on the Lerwick rubbish-dump (on the site of the North Loch).

*C. nigra. *Common Knapweed.* A local and sometimes long-persistent casual of roadsides and grassy places, occasionally well away from houses; also by the edges of cultivated fields.

*Cichorium sp. *Chicory*. One plant, probably *C. intybus*, occurred as a casual in the garden at 'Lochside', North Roe, Northmavine (2002, A. Williamson).

*Lapsana communis subsp. **communis**. *Nipplewort*. A local casual. Primarily a weed of gardens and (rarely) of rough grassy or waste places. It is even rarer as a weed of arable ground, as in a fallow field at Easter Quarff, South Mainland (1959), and two years later in cornfields in the Sound area of Weisdale, Central Mainland. It can persist for many years where conditions remain suitable.

*Hypochaeris radicata. *Cat's-ear*. A well-established alien of dry places in and near inhabited areas: grassy roadsides and embankments, pastures, and low grassy sea-banks. First certainly recorded from Shetland, in 1889, by W. H. Beeby in these words, 'While driving into Lerwick I detected a few plants on a dry, grassy bank, on the S side of Dales Voe growing amid a profusion of the ubiquitous *Leontodon autumnalis.*' It may have been in Shetland much earlier if J. Robertson's unlocalised report of it is to be accepted.

Leontodon autumnalis. *Autumn Hawkbit*. A very common and widespread native which is found in a wide range of (mainly) grassy habitats: in dry or wet pastures (especially short coastal turf and saltmarshes); meadows; on turfy, grassy heaths; bare granite fellfield; and on roadsides and waste ground.

*L. hispidus. *Rough Hawkbit*. A very rare casual. One plant on an embankment by the Mid Yell Junior High School (1997, A. D. D. Gear). Not seen again.

*Picris echioides. *Bristly Oxtongue*. Another very rare casual. A number of plants on agricultural ground at Brae and Ham, both in Foula (1999–2001, S. Gear).

*Sonchus arvensis. *Perennial Sow-thistle*. This is a well-established alien of arable ground; on the upper parts of beaches among boulders or on shingle or sand; and on foreshore walls and banks in the inhabited areas. In the past it often occurred in cornfields, a habitat rarely seen nowadays. On some beaches it is notably abundant, as around the head of Harold's Wick, Unst. A form with glabrous involucres and peduncles grows around the school and schoolhouse in Fair Isle from where it has been known for a very long time. In 2004 WS counted 250 plants by the outer side of the surrounding walls and by the playground area, a figure which excludes significant numbers within the walls. Another form with glabrous peduncles, but with glandular-hairy involucres, was recorded in 1977 from the N side of the Pool of Virkie, South Mainland.

*S. oleraceus. *Smooth Sow-thistle*. Casual. Primarily a weed of gardens; also on rubbish-dumps and other open disturbed areas, and by walls, etc. Frequent in Lerwick and Scalloway and other smaller centres of population, occasional elsewhere.

*S. asper. *Prickly Sow-thistle*. A well-established alien which is occasionally seen in cultivated ground (especially cornfields); on heaps of rubbly soil and on rubbish-dumps; and by roadsides and as a weed in gardens. Formerly more prevalent when cornfields were distributed throughout the county.

*Lactuca sativa. *Garden Lettuce*. Casual. Recorded in 1967 from the Lerwick rubbish-dump (on the site of the North Loch), and in 1968 from the Scalloway rubbish-dump. A mere garden outcast in both cases.

*Cicerbita macrophylla subsp. **uralensis**. *Common Blue-sow-thistle*. Casual, rarely seen as a garden escape or outcast. It is capable of persisting, a good example being near the mouth of the Burn of Kirkhouse, Voe, Central Mainland, where it grew from 1963 or earlier to 1982 when it was recorded as established. It was next searched for in 1999 (R. C. Palmer) and 2008 (WS), but could not be found.

*Mycelis muralis. *Wall Lettuce*. Casual, once recorded. One plant as a weed in a garden at Quoyness, above the Loch of Strom, Central Mainland (1996, J. Clark).

Taraxacum. *Dandelions*. Apart from the eight native or probably native species, and which are treated in the normal way, all of the others—a total of fifty-five—are regarded as well-established aliens despite the fact that many are known from very few stations. The writer feels that this is a better placement than listing them as casual aliens because of their perennial nature, abundant seed-set, and their ability to survive and spread where conditions remain suitable. The well-established aliens are not usually provided with habitat details. This is because they are all inhabitants of man-made sites, or of sites influenced in some way by man: by roadsides and ditch banks; in or by cemeteries, churchyards and schoolyards; on and by old walls and in disused quarries; in gardens, lawns, and fallow fields; on open waste rubbly ground; and on low sea-banks, foreshores and sand-dunes in the inhabited areas. In order to avoid repetition of habitats, the reader may assume that the alien dandelions have all occurred in at least one of the habitats listed above. Localities for the rare aliens are given briefly; otherwise a general note regarding their frequency and distribution is provided for the less rare taxa. The vast majority of our *Taraxacum* records were made from 1974 to 1980, making a new survey of the genus (particularly of the alien species) overdue. Recent and inevitable changes in modern *Taraxacum* taxonomy, along with the renaming of some Shetland material, have led to a few differences between this brief synopsis and the the rather more detailed account of the genus in Scott & Palmer (1987).

Section Erythrosperma.

T. brachyglossum. *Purple-bracted Dandelion*. Probably native. Sandy turf by the West Pool, Mousa, off South Mainland.

*__T. scanicum__. *Skåne Dandelion*. Sumburgh Airport, and the Links of Quendale, both in South Mainland; East Voe of Scalloway.

*__T. oxoniense__. *Oxford Dandelion*. Near the pier, Mid Yell.

Section Spectabilia.

T. faeroense. *Faeroes Dandelion*. The most common and widespread dandelion in Shetland and tolerant of both dry or wet places: by rocky-heathery burnsides; grassy niches among rocky outcrops; marshy pastures; turfy spots on heaths and moors; stony loch margins; and in man-made habitats such as roadsides, ditch banks, etc.

†**T. geirhildae**. *Shetland Dandelion*. Endemic. Grassy ledges and niches among rocks. This was first recorded, in 1907, by W. H. Beeby from near Lang Clodie Loch, North Roe, Northmavine, and where it grows to this day in the Lang Clodie Loch and Birka Water areas. In 1993 it was found in a few places on the N side of Ronas Voe and—for the first time outside Northmavine—near Kellister, West Mainland. Two more records came to light in 2006, both made by WS: very sparingly near the Round Loch, near Grut Wick, Lunna Ness, North Mainland; and a small colony on a crag at Hurda Field, N of Mavis Grind, Northmavine.

†**T. serpenticola**. *Serpentine Dandelion*. Endemic. Known only from around the summit of Muckle Heog, near Baltasound, Unst, in grassy places among serpentinite outcrops.

T. 'Taxon 2177'. This reference has been assigned by WS to an apparently native member of this section which R. C. Palmer found in 1974 on rocks by the lower part of the Burn of Laxdale, Cunningsburgh, South Mainland (reference no. 2177 in the writer's herbarium). The following year Palmer said that it had been exterminated by gravel dumping, but in 2007—the first time that it had been searched for after its apparent disappearance—WS came across a small healthy colony at the original site. (Any gravel which had earlier covered it had presumably been washed away by the burn in spate, leaving the roots intact in rock-crevices.) Although it is new to Britain and may be an undescribed species, it is also equally likely that it has close relatives in Iceland or Faeroe if not conspecific with a taxon from these areas. Further studies are needed to assess its taxonomic position.

Section Naevosa.

*T. naevosum. *Squat Dandelion*. Frequent about Lerwick and Scalloway, very rare elsewhere.

*T. naevosiforme. *Wetland Dandelion*. Lerwick, in a few places, and (single records only) from Gulberwick (S of Lerwick) and in Scalloway.

*T. euryphyllum. *Wide-stalked Dandelion*. This is one of our more widespread alien dandelions. Occasional in South Mainland, and in Central Mainland where it becomes frequent in and around Scalloway and Lerwick. Very rare in West Mainland and Northmavine. It is also recorded from Fair Isle, Bressay, Yell (especially about Burravoe), and from one place in Unst.

†T. hirsutissimum. *Hairy Dandelion*. Endemic. Known only from dry grassy and often sandy pastures by the coast, sand-dunes, and roadside verges in the S of South Mainland from near Williamsetter and S to Scousburgh (including St Ninian's Isle), and from the Links of Quendale to Grutness. There is one outlier farther N, near South Voxter, Cunningsburgh. It is rare in some areas, but more frequent in others.

*T. maculosum. *Spotted Dandelion*. Three records only: Walls, West Mainland; Hillswick, Northmavine; West Sandwick, Yell.

*T. pseudolarssonii. *Spreading-bracted Dandelion*. Wick of Sandsayre, Sandwick, South Mainland; Walls, West Mainland.

*T. subnaevosum. *Pale-bracted Dandelion*. Gremista, near Lerwick; Uphouse, Bressay.

Section Celtica.

*T. gelertii. *Gelert's Dandelion*. Known only from Tingwall churchyard, Central Mainland.

*T. bracteatum. *Dark-green Dandelion*. Known only from Fair Isle.

*T. duplidentifrons. *Double-toothed Dandelion*. This is probably our commonest and most widespread alien dandelion, and is recorded from Fair Isle to Unst.

*T. inane. *Pollenless Dandelion*. Symbister, Whalsay; N of Baltasound, Unst.

*T. landmarkii. *Landmark Dandelion*. Near the jetty by the W side of the Loch of Tingwall, Central Mainland; West Sand Wick, Yell.

*T. nordstedtii. *Nordstedt's Dandelion*. Near Asta, N of Scalloway; Fort Charlotte, Lerwick.

T. fulvicarpum. *Brown-fruited Dandelion*. Probably native. Recorded from South Mainland only. On sand-dunes at the S end of the Bay of Scousburgh, and Leven Wick; and (in 1953) in a quarry between Bakkasetter and Symblisetter.

T. unguilobum. *Claw-lobed Dandelion*. Probably native, at least in South Mainland. Dry grassy places: among sand-dunes or on sandy soil; in turf among rocky outcrops; streamsides; and on the verges of roads and paths. Not uncommon in South Mainland (especially in and near Cunningsburgh); very rare in Central and North Mainland, Northmavine, Bressay and Yell; uncommon in Unst, and not on record from West Mainland.

*T. luteum. *Pure Yellow Dandelion*. Known only from Hillswick, Northmavine.

Section <u>Hamata</u>.

*T. **hamatum**. *Hook-lobed Dandelion.* Rare or occasional throughout the county, and extending to Yell and Unst.

*T. **hamatulum**. *Slender Hook-lobed Dandelion.* Fair Isle; Scalloway, and nearby Asta; Lerwick.

*T. **subhamatum**. *Large Hook-lobed Dandelion.* Roadside by the Loch of Tingwall, Central Mainland; Lerwick.

*T. **quadrans**. *Fleshy-lobed Dandelion.* Lerwick; Haroldswick, Unst.

*T. **pseudohamatum**. *False Hook-lobed Dandelion.* Sandness, West Mainland.

*T. **atactum**. *Narrow-bracted Dandelion.* Lerwick; Voe of Cullingsburgh, Bressay; Sandness, West Mainland; Baltasound, Unst.

*T. **hamatiforme**. *Asymmetrical Hook-lobed Dandelion.* There are single records from South, West, and North Mainland, Fetlar, Unst, and two from Central Mainland.

*T. **lamprophyllum**. *Lustrous-leaved Dandelion.* Burravoe, Brae, North Mainland.

Section <u>Ruderalia</u>.

*T. **macrolobum**. *Incise-lobed Dandelion.* Pool of Virkie, South Mainland.

*T. **pannucium**. *Green-stalked Dandelion.* Four records, two in the Sumburgh area, South Mainland, and two in the Scalloway area.

*T. **tenebricans**. *Shiny-leaved Dandelion.* West Voe of Sumburgh, South Mainland; Muckle Roe.

*T. **dilaceratum**. *Lacerate-leaved Dandelion.* Scalloway, and nearby Asta; Bixter, West Mainland.

*T. **alatum**. *Green Dandelion.* Lerwick; Walls, West Mainland.

*T. **insigne**. *Remarkable Dandelion.* Levenwick, South Mainland; Fort Charlotte, Lerwick; near the Voe of Leiraness, Bressay.

*T. **laticordatum**. *Decumbent Dandelion.* Near the quay, Hamnavoe, West Burra; Hillswick, Northmavine.

*T. **pallescens**. *Pink-stalked Dandelion.* Walls, West Mainland.

*T. **expallidiforme**. *Broad-stalked Dandelion.* Sumburgh Airport, South Mainland.

*T. **croceiflorum**. *Orange-flowered Dandelion.* Near Scalloway Castle.

*T. **cyanolepis**. *Bluish-bracted Dandelion.* Roadside, W side of Lax Firth, Central Mainland.

*T. **ancistrolobum**. *Few-lobed Dandelion.* Lu Ness, S of Hamnavoe, West Burra; Scalloway; Tingwall churchyard, Central Mainland; Haroldswick, Unst.

*T. **sellandii**. *Selland's Dandelion.* Near Bakkasetter, South Mainland; Scalloway; 'Gardie House', Bressay; Mid Yell.

*T. **altissimum**. *Tall Dandelion.* West Voe of Sumburgh, South Mainland.

*T. **adiantifrons**. *Pretty-leaved Dandelion.* Sumburgh Airport, South Mainland.

*T. **acroglossum**. *Broad-bracted Dandelion.* Near Sandwick, Whalsay.

*T. **vastisectum**. *Crowded-lobed Dandelion.* Lerwick; West Voe of Sumburgh, South Mainland.

*T. **cordatum**. *Entire-lobed Dandelion.* Scalloway; Brough, South Nesting, Central Mainland.

*T. **ekmanii**. *Ekman's Dandelion.* There are five records, one each in South Mainland, Central Mainland, West Mainland, Fetlar, and Unst.

*T. **oblongatum**. *Oblong-leaved Dandelion.* Hillswick, Northmavine.

*T. **tanyphyllum**. *Spreading-lobed Dandelion.* Levenwick, South Mainland.

*T. **dahlstedtii**. *Dahlstedt's Dandelion.* Six records: two in South Mainland, two in Central Mainland, and one each in Bressay and Northmavine.

*T. **latisectum**. *Broad-lobed Dandelion.* Ollaberry, Northmavine.

*T. **subundulatum**. *Complex-leaved Dandelion.* Symbister, Whalsay.

*T. **pectinatiforme**. *Pectinate-leaved Dandelion.* Hoversta, Bressay.

*T. **caloschistum**. *Brilliant-stalked Dandelion.* Sandwick, South Mainland.

*T. **polyodon**. *Common Dandelion.* Recorded from Fair Isle; Sumburgh area, South Mainland; Scalloway; Bressay; Brae, North Mainland; and Ollaberry, Northmavine.

*T. **incisum**. *Incise-leaved Dandelion.* Baltasound, Unst.

*T. **xanthostigma**. *Ochre-styled Dandelion.* Sandness, West Mainland.

*T. **longisquameum**. *Elongate-bracted Dandelion.* Lerwick.

*T. **scotiniforme**. *Deltoid-lobed Dandelion.* Lerwick.

*T. **fasciatum**. *Dense-bracted Dandelion.* Levenwick, South Mainland; Scalloway; near the W side of the Loch of Tingwall, Central Mainland; Lerwick.

*__Crepis biennis__. *Rough Hawk's-beard.* A very rare casual, once recorded. A single plant on the reseeded embankment of the (old) A970 just above Channerwick, South Mainland (1999, A. Newton).

*C. **tectorum**. *Narrow-leaved Hawk's-beard.* A very rare casual with three records only, all of single plants: cornfield, Bakkasetter, South Mainland (1980); sown pasture near the SE corner of the Loch of Clickimin, Lerwick (1982); new roadside verge by Scalloway Castle (1983).

*C. **capillaris**. *Smooth Hawk's-beard.* A rare casual of arable and waste ground, gardens, etc., and with scattered records from South, Central, and West Mainland, and Northmavine. The latest record is as follows: 'quite a strong colony' on a grassy roadside bank, entrance to 'Burrastow House', SW of Walls, West Mainland (1987, R. W. M. Corner).

*C. **setosa**. *Bristly Hawk's-beard.* Another very rare casual. One plant in sown pasture near the SE corner of the Loch of Clickimin, Lerwick (1982), and three plants on waste ground on the S side of the Unst Leisure Centre (1989, R. C. Palmer).

†**Pilosella flagellaris** subsp. **bicapitata**. *Shetland Mouse-ear-hawkweed.* Endemic. Dry heathy, often rocky, pastures, and steep rocky sea-banks. Known from three sites only. White Ness peninsula, Central Mainland, sparingly, but spread over a wide area; N side of West Burra Firth, West Mainland; and on steep sea-banks in The Kirk area , N side of Ronas Voe, Northmavine. First discovered, in 1962, at White Ness. It flowers very sparingly at White Ness, but a little more freely at the other places where grazing pressure may be less severe.

*P. **aurantiaca**. *Fox-and-cubs.* Casual. Fairly often grown in gardens and not infrequently seen outside them on grassy banks and roadside verges as an escape or outcast; perhaps sometimes a deliberate introduction. It is capable of persisting and spreading where conditions remain suitable.

Hieracium. *Hawkweeds.* Twenty-seven taxa are listed here, including 'Taxon G' noted under *H. sparsifolium* below. Eighteen of these are considered endemic to the county and embrace all sixteen members of Section Alpestria, with an additional two from other sections. Only two of the twenty-seven species are believed to be extinct: *H. maritimum* and *H. hethlandiae*, the latter being also one of our endemics. Between

64

1999 and 2005 Scottish Natural Heritage introduced cultivated material of our rarest endemic hawkweeds into selected areas in the wild to safeguard against their possible total loss to science. The species chosen for this trial scheme were *H. amaurostictum, H. attenuatifolium, H. breve, H. difficile,* and *H. hethlandiae.* Unfortunately, results were disappointing except in the case of *H. attenuatifolium* where they were more encouraging. The trial was replaced by a dedicated horticultural unit near Lerwick where all of our endemic Hieracia are being successfully grown. The writer's own attempts to introduce cultivated hawkweeds into the wild also met with little success.

Section Foliosa.

[**H. maritimum**. *Maritime Hawkweed.* Believed to be extinct. This was first and last seen in 1902 by W. H. Beeby on a holm in the N end of Burga Water, between Walls and Sandness, West Mainland. Several attempts have been made to refind it but all to no avail. It seems likely that it was crowded out by *Luzula sylvatica* during the first half of the twentieth century.]

Section Tridentata.

H. gothicoides. *Broad-headed Hawkweed.* Very rare and found on rocky outcrops and streambanks in only three places, all in North Mainland, and becoming rarer if not nearing extinction in the first two stations. Ravine of the Mill Burn, near Swining; first found here in 1960 but only one plant was seen in 1997, by L. A. Inkster. Steep rocks W of Lunning (1962); for some years this remained a good station but in 2004 P. V. Harvey, J. Swale, and WS saw just one plant in bud and nine seedlings. In 1980 a few plants were noted in the Fossnis Dale and Dallican Water areas at Catta Ness, NW of Lunning; in 2004 the same recorders found a small but healthy colony in this area containing twenty plants in flower and, nearby, at least fifty seedlings.

H. lissolepium. *Hairless-bracted Hawkweed.* Extremely rare. Found in 1961 on the steep heavily vegetated side of a *geo* near Eric's Ham, NE of Aywick, Yell. Because of an unfortunate slip, the locality was wrongly described by Scott & Palmer (1987). This led to more than one unsuccessful attempt to refind it, and to the belief for many years that it become extinct. Fortunately, in 1988 WS located the original station. In 2004 P. V. Harvey counted ten stems in flower at this almost inaccessible site.

H. sparsifolium. *Sparse-leaved Hawkweed.* Rocky outcrops, sea-banks, streamsides, pastures, and (rarely) on rocky banks by lochs. Local in Central and West Mainland, Yell, Fetlar, and Unst, rare in West Mainland and unrecorded from South Mainland. The leaves are nearly always heavily purple-blotched; colonies with leaves wholly green are very rare. Included here, and in the statistics of this checklist, is a form noted in 1988 by J. Bevan and referred to by WS as 'Taxon G'. It differs slightly from the usual Shetland plant but may not warrant taxonomic distinction.

Section Alpestria.

†**H. vinicaule**. *Wine-stemmed Hawkweed.* Endemic. Sea-banks and rocky burnsides; rarely in pastures, on rocky outcrops, holms in lochs, or by roadsides. Local to frequent in North Mainland and Northmavine, rare in West Mainland and Yell. The following new discoveries have been made: on several crags on the N side of Gunnister Voe, Northmavine, about forty stems in total (1991, L. A. Inkster); about fifteen stems among much greenery (*Luzula sylvatica, Solidago virgaurea,* etc.) on the steep sea-bank by the E side of Orr Wick, head of Ronas Voe, Northmavine (2006, WS).

†**H. northroense**. *North Roe Hawkweed.* Endemic. Very rare. Known only from a grassy rocky meadow and a small rocky chasm in the Burravoe area of North Roe, Northmavine. In 1994 there were *c.*800 plants at the former site and (in 2000) about eighteen in the latter station. Two small colonies, totalling about sixty plants in 1992, occur on steep sea-banks at the SW side of the Voe of Snarraness, West Mainland. Its official protection at the North Roe meadow site is now (2010) a pretence.

†**H. klingrahoolense**. *Klingrahool Hawkweed.* Endemic. Local, rarely abundant. Rocky or grassy sides of streams, rocky outcrops, sea-banks, meadows and rocky pastures, and (rarely) on holms in lochs. In Vementry island it occurs in two or possibly three places, and is abundant in one of these sites. There are two stations in Central Mainland, three in North Mainland (including an outlier at Kels Wick, Lunna Ness), and two in Yell (including a holm in the Loch of Lumbister). However, its best site by far is the E side of Muckle Roe, on low sea-banks and in adjacent meadows and rocky pastures to the NW and S of the bridge. Here it grows abundantly, along with less common *H.vinicaule*, and scarce *H. sparsifolium* and *H. subtruncatum*. Its name, of Norse origin, commemorates a place in South Nesting, Central Mainland, near where it was first recorded. Formerly listed as 'Taxon B', 'Taxon C', or 'Taxon B/C'.

†**H. subtruncatum**. *Mainland Hawkweed.* Endemic. Low grassy sea-banks, rocky or grassy burnsides, rocky outcrops, and (rarely) in meadows and pastures. Very rare in Central Mainland where it is probably nearing extinction in two of its three sites (see below). Rare in West and North Mainland. In Muckle Roe there are five sites, and in Northmavine it is occasional from Islesburgh, near Mavis Grind, N to the N side of Ronas Voe. In 2003 severe floods at the Burn of Channerwick exterminated a fine colony, the only known site for this species in this division. It formerly flourished in meadows and pastures at Tresta (near Bixter) and at Garth (South Nesting), both in Central Mainland. In 1999 R. C. Palmer found none at Tresta; in 1984 he saw about twenty plants at Garth. The writer, in 1963, noted it from the 'west side of Hamna Dale, and northwards to near Lunning', North Mainland; in 2006 he saw none at all. Three new records exist, and are included in the above summary: E side of Fora Ness, entrance to Sandsound Voe, West Mainland, frequent over a large stretch of sea-bank (1987, WS); *c.*150 plants on a sea-bank N of Jackville, W side of Stromness Voe, Central Mainland (1993, WS); on sea-banks at the mouth of the Loch of Scarvister, near Skelda Ness, West Mainland (1994, L. A. Inkster).

†**H. dilectum**. *Purple-tinted Hawkweed.* Endemic. Very rare. Rocky outcrops and streamsides. This is known from four sites: on rocks above the W side of the Bay of Brenwell, West Mainland; about seventy stems on rocks, Muckle Hoo Field, NE of West Burrafirth, West Mainland (1998, WS); at the lower end of the Laxo Burn, Central Mainland, fifty-five plants (2006, J. Swale); two colonies by the Burn of Quoys, South Nesting, Central Mainland, the lower with only a mere four plants (2006, WS).

†**H. pugsleyi**. *Pugsley's Hawkweed.* Endemic. Very rare. Steep grassy or rocky sea-cliffs and sea-banks; in ravines; by burns and (very rarely) roadsides. Known from two sites in Yell, and one each in West, Central, and South Mainland. In Yell there are hundreds if not thousands of plants on sea-banks by the W side of Whale Firth, from Birka Lees to North Fiski Geo. It also grows in the ravine running into Skurdie

Geo, Otters Wick, Yell , where 186 plants were counted in 1987 by J. Bevan, and extends to the adjacent and very steep sea-cliff. In 1987 JB noted forty-five plants by or near the roadside at Tumblin, Bixter, West Mainland (HU339531); in 1996 WS could find none. The Central Mainland site is at the lower end of the Burn of Weisdale, near the old mill (now the Bonhoga Gallery); in 1987 JB counted forty plants, but in 2006 only eight examples were seen (BSBI field meeting). In South Mainland it occurs or has occurred in many places by the Burn of Laxdale, Cunningsburgh, from just below the A970 to near its mouth. In 2005 J. Swale found 194 adult plants from the A970 down to the Tow–Gord lane; the remainder of the burn, however, needs a new survey. The Skurdie Geo and Whale Firth plants were formerly referred to as 'Taxon D' and 'Taxon F' respectively.

†**H. spenceanum**. *Spence's Hawkweed*. Endemic. Very rare. Rocky turfy heaths; on crags and sea-banks; and by burns and roadsides. Confined to the northern portion of West Mainland from the Norby–Bousta areas of Sandness E to the West Burrafirth–Neeans–Brindister district. This hawkweed has suffered badly from overgrazing and in some sites it may now be represented by tiny barren rosettes. There is, fortunately, one place where it is in fine condition: West Burrafirth, on low sea-banks between the jetty and the head of the *voe*, at least 300 stems and numerous rosettes (1992, L. A. Inkster and WS). In 1993 at the nearby Ward of Scollan WS saw between sixty and eighty plants on a crag; in 2006 he found only seventeen flowering stems and fifteen rosettes on the same rock. Formerly referred to as 'Taxon A'.

†**H. attenuatifolium**. *Laxo Burn Hawkweed*. Endemic. Extremely rare. Known only from grassy or rocky places at or near the mouth of the Laxo Burn, E of Voe, Central Mainland. Repeated introductions here over several years have had a very low survival rate in spite of the fact that it is a robust grower in cultivation. In 2006 J. Swale counted fourteen flowering plants and ten seedlings. The grazing management measures, which were put in place at the Laxo Burn specifically to assist the recovery of *H. attenuatifolium*, have been of more benefit to *H. dilectum*; its numbers have slightly increased while those of the present species have remained disappointingly static. Its official protection is now (2010) a pretence.

[†**H. hethlandiae**. *Cliva Hill Hawkweed*. Believed to be an extinct endemic. This survives in cultivation, but is not now known to occur in the wild except for a possible plant or two in an area where it has been deliberately introduced. It grew in just one station: the steep rocky-heathery side of Cliva Hill, by the A970 next to Mavis Grind, North Mainland. (A claim that it was found at Ronas Voe, Northmavine, in 1920 has never been confirmed.) In 1976 the Cliva Hill habitat was completely destroyed by quarrying, and *H. hethlandiae* and other scarce hawkweeds exterminated. Prior to this, WS removed seven plants of the Cliva Hill rarity, thus ensuring the existence of this very rare plant in cultivation, if not in the wild. In 1980 about seventy-five plants and hundreds of achenes were introduced among rocks at Hurda Field, N of Mavis Grind, Northmavine. A further twelve plants were added a year later. In 1982 no seedlings could be found, and only a handful of plants had survived. This led to the experiment being terminated in 1983 when the remaining plants were removed and brought back into cultivation. Two more introductions were attempted: in 1987 near the lower end of the Burn of Skelladale, near Brae, North Mainland, and in 2001 at the E side of Muckle Roe. Both have been met with very little success.]

†**H. praethulense**. *Thule Hawkweed.* Endemic. Very rare. Rocky sea-banks, crags, and holms in lochs. All of its recorded stations except one are in Northmavine where its stronghold lies in the Ronas Voe and Ronas Hill areas. The exception is in North Mainland. In Northmavine one site exists by the S side of Ronas Voe, at HU323806; on the N side of the *voe* there are numerous stations from the Pobie Sukka area to The Kirk area, on cliffy sea-banks and among the crags above at various altitudes. Its best site is at Swabie Water, N of Ronas Hill where it is abundant on the two adjacent holms in the loch. In 2005 T. Harrison found a mature plant by the W side of Swabie Water. There are also records from three places at the S end of Northmavine, but in all cases the plant has fared badly. These are: by the NE arm of Mangaster Voe, last seen in 1987 (WS); on rocks by a lochan SW of the Loch of Haggrister, in poor condition in 2000 (WS); and on and below rocks by the main road near Innbanks, Mangaster Voe, also in poor condition in 2000 (WS). The North Mainland record refers to Cliva Hill, by Mavis Grind, where it was collected in the early 1920s; no later note exists from here, and in 1976 the site was destroyed by quarrying. *H. praethulense* was accidentally introduced in 2001 in the Burn of the Twa-roes area on the NE side of Ronas Hill, Northmavine.

†**H. australius**. *Unst Hawkweed.* Endemic. Very rare, but relatively frequent in the N of Unst. Confined to Unst, Fetlar, and Northmavine. In Unst it grows by the SE, NW and NE sides of the Loch of Cliff. It also grew by the N side of the arm of the Loch of Cliff, as well as by the Burn of Burrafirth, and on the Burra Firth cliffs, but has not been seen in these areas since 1962, 1959, and 1950 respectively. In Fetlar it is found in one place: on steep rocky sea-banks in the Red Geo area of the Wick of Tresta. Its only currently known Mainland sites are in the Ronas Voe and Ronas Hill areas of Northmavine: at Nevi Geo, by the S side of the *voe* (1992, WS), and in a number of places on the N side from The Black Well area to The Kirk area, from near sea-level to high up among the crags. It was recorded in the early 1920s from Cliva Hill, by Mavis Grind, North Mainland; there is no later record from here, and in 1976 the site was destroyed by quarrying.

†**H. difficile**. *Okraquoy Hawkweed.* Endemic. Extremely rare. This taxon is known from one site, the narrow limestone ravine running into the Bay of Okraquoy, between Fladdabister and Aithsetter, South Mainland. It is now the most southerly site in Shetland for a hawkweed, following the removal by flooding of *H. subtruncatum* from the Burn of Channerwick. Estimates of the number of plants at Okraquoy vary from *c.*100 to twice that figure. A large number of plants were introduced to a limestone quarry near Voe, Central Mainland, in 2001–2002; by 2004 only a handful had survived, and in 2009 L. A. Inkster and WS noted about fifteen plants only, a few of which were in bud.

†**H. amaurostictum**. *Semblister Hawkweed.* Endemic. Extremely rare. Low grassy rocky sea-banks E of the Loch of Semblister, at the entrance to The Firth, near Bixter, West Mainland. In 1993 WS noticed at least 150 stems and numerous seedlings. More recent observers have recorded much lower numbers of stems while still noting the abundance of seedlings. Perhaps the number of adult plants varies a great deal from one season to another. Formerly referred to as 'Taxon E'. In 2005 *c.*150 plants were introduced to a site at Tresta, near Bixter, Central Mainland. By 2008 hardly any had survived.

†**H. gratum**. *Handsome Hawkweed.* Endemic. Very rare. Steep dry grassy (often rocky or heathery) sea-banks, and rocky banks by lochs. Known only from Yell and Unst. In 1865 it was discovered at Burra Firth, Unst, by R. Tate, and for well over a century it remained an Unst speciality until 1988 when J. Bevan came across it in the adjacent island of Yell. It grows in the following four areas about the head of Burra Firth: Fiska Wick, about eighty plants (1988, J. Bevan); *c.*240 plants (including many rosettes) from just S of the Boo Stacks to above the pool at the mouth of the Burn of Burrafirth (1988, JB, also 1992, L. A. Inkster and WS); Stabba, eighty-six plants and twenty seedlings (1988, JB); fifteen small plants on the sea-bank below Buddabrake, just S of the tiny stream (1990, WS, also 1992, LAI and WS). In 1886 W. H. Beeby found it in abundance by the arm of the Loch of Cliff; by 1962 it had disappeared. However, it was 'doing well' by the NE corner of the Loch of Cliff, N of the entrance to the arm (1992, WS). In Yell it grows in the following three places, all by the W side of Whale Firth: twenty-two plants on a low sea-bank at HU464938, NNE of North Grommond; small amounts in North Fiski Geo; and (sparingly) high up in Longi Geo. This is now the most northerly endemic plant in Britain, a distinction it used to share with *H. australius* until that species apparently disappeared from Burra Firth.

†**H. breve**. *Rare Hawkweed.* Endemic. Exceedingly rare. Rocky granite outcrops above the N side of Ronas Voe, Northmavine. This was first discovered by W. H. Beeby in 1889 in HU3081 where there may be no more than a score or so of plants. Tiny flowerless plants of *H. praethulense* (which is frequent in the area) grow on the same rocks and can look very like *H. breve* in the same condition. This has given rise to overly high numbers of the rarity. A second and much smaller colony was found in 1992 by L. A. Inkster and WS (HU2981); here, there are only a handful of stems. The new site is probably inaccessible to sheep; the same is not true of Beeby's station. There may be other sites for *H. breve* waiting to be discovered in this very large area which abounds in apparently suitable habitats. In 2001 and 2005 a total of well over 200 plants were introduced in the Burn of the Twa-roes area on the NE side of Ronas Hill, Northmavine. Some of the material from the earlier year turned out to be *H. praethulense*, having been inadvertently included with *H.breve*.

†**H. zetlandicum**. *Shetland Hawkweed.* Endemic. Rare. Rocky or grassy sea-cliffs and sea-banks; grassy, heathery or rocky burnsides; and (now very rarely) in grassy rocky meadows. This is another of W. H. Beeby's fine discoveries. Until it was found in the 1960s in West and Central Mainland, *H. zetlandicum* was known only from an area of hilly pastures in North Roe, Northmavine, from Burgo Taing (near Burravoe) to Benigarth (near Isbister). Within this area it grew plentifully, but in recent decades it has become confined to two small spots within it, namely, a grassy rocky meadow below Burravoe (about forty plants, 1994, A. J. Kerr and M. B. Usher), and ten plants by the grassy banks of a small burn above Isbister (1992, L. A. Inkster and WS). In 1924 it was observed on the Lee of Setter, N of Isbister. Elsewhere in Northmavine it grows in the following four coastal sites: very sparingly on the S side of Ronas Voe (HU3280), and scattered specimens between Troti Geo and the Point of Quida-stack, also on the S side; a small colony at The Neap, NE of Burravoe, and twenty plants and a number of rosettes at Eislin Geo, near Fethaland. The Central Mainland site, at the Burn of Lunklet, is the finest; here, in 1999, grew 334 adults and

367 young plants and seedlings (P. V. Harvey). Finally, there are two records from West Mainland: heathy pasture, W side of Snarra Ness, first seen in 1963 but not noted by later searchers; rocky *geos*, SW side of the Voe of Snarraness (1964), twenty-five plants in flower along with a good number of seedlings (2004, PVH).

Section Oreadea.

H. subscoticum. *Tapered-leaved Hawkweed.* Endemic. Very rare and confined to Northmavine. This is now known only from a section of steep craggy sea-cliffs, inshore stacks, and scree-filled coastal gullies on the N side of Ronas Voe from just NW of Slocka to the waterfall SE of The Kirk. Here, where most of its sites are inaccessible, it is common and probably locally abundant as far as can be ascertained by the use of a boat and binoculars. In 1963, when it was first discovered, it grew in pastures and by streams at Heylor, by the S side of Ronas Voe. It was last seen there the following year. Fortunately, in 1967 the more secure sites on the other side of the *voe* were discovered. Formerly referred to as '*H. scoticum* North Mainland'.

†H. scottii. *Scott's Hawkweed.* Endemic. Very rare and confined to West Mainland where there are two stations. First discovered in 1964 in rocky *geos* at the SW side of the Voe of Snarraness. In 2004 P. V. Harvey counted twenty plants in flower along with a good number of seedlings. Not far away, at Bousta, Sandness, in 1967, it grew in pastures and on associated rocky outcrops. In 1988 WS observed it on rocks immediately W of Bousta, and towards Ness. Since then overgrazing has probably caused a decline in the vigour and number of plants. An intensive survey of the Bousta area is required. At the Voe of Snarraness it grows with *H. zetlandicum*, and at Bousta by *H. caledonicum*. Formerly referred to as '*H. scoticum* West Mainland'.

H. beebyanum. *Beeby's Hawkweed.* In crevices or grassy niches among crags and rocky outcrops; on steep rocky or grassy-heathery sides of burns; and rarely on dry banks and heathy pastures. This hawkweed grows in Mainland, Muckle Roe, and Vementry. In Mainland it is frequent in the N part of West Mainland, and also in Northmavine, especially in the Ronas Voe and North Roe areas. Outliers are found at the Burn of Valayre, near Brae; Lunna Ness; and S of Lunning, all in North Mainland; and W of Gletness, Central Mainland. Its most southerly occurence is in Central Mainland—a tiny patch among limestone rocks on the White Ness peninsula. The writer, after visiting nearly all the known sites for this plant in 2006, is of the opinion that the bulk, if not all, of the material comes under *H. beebyanum.* He is aware that the closely related *H. orimeles* is also on record from a very few places in Shetland, but is reluctant to accept it until its presence has been conclusively proved.

Section Stelligera.

H. caledonicum. *Caledonian Hawkweed.* Rare. Rocky outcrops and steep, often craggy sea-banks and sea-cliffs. Known only from the northern coastal fringe of West Mainland, and from North Mainland and Northmavine. In West Mainland it grows from Bousta, Sandness (over fifty flowering stems (2004, P. V. Harvey)), to the entrance to Aith Voe, where in 1992 L. A. Inkster and WS found it plentifully on sea-banks to the E and W of Keen Point. Within this area it occurs by the W side of the Stead of Aithsness, as well as about Maa Loch in the island of Vementry. One new record exists for this district: two plants in flower and twenty-five rosettes in severely grazed rocky pasture just west of the rock-face (with planted shrubs) by the road at Vementry (Mainland) (2006, WS). In North Mainland it is found on steep rocky-heathery sea-banks at the Ness of Houll, SW of Mavis Grind (1984, WS). It occurs plentifully (with *H. subscoticum* and other hawkweeds) on rocky sea-cliffs and crags, inshore stacks, and in scree-filled coastal gullies from just NW of Slocka to near The Priest, on the N side of Ronas Voe, Northmavine

H. argenteum. *Silvery Hawkweed.* Rare. Rocky outcrops and rocky heathy hillocks. Known only from the West Burrafirth district of West Mainland, Muckle Roe, and the Ronas Voe and North Roe areas of Northmavine. At West Burrafirth it is found on rocks above the S side of Longa Water and in the gully between the loch and the sea, also, but very sparingly, on a small rocky hillock NE of the broch. The Muckle Roe station is a recent discovery: on a line of SW-facing crags 250 m NNE of the N corner of the North Ham, in the NW of the island (2001, A. K. Thorne). In Northmavine it is recorded from a gully at Turls Head and on a rocky face at the Moshella Lochs, both sites NW and N of Ronas Hill respectively. There are two records from above the N shore of Ronas Voe: two rosettes on the edge of The Trip, some way above The Shun (1992, 1994, L. A. Inkster and WS); The Brough, perhaps a score or more of plants (some in flower) to the N of the source of the N fork of the Y-shaped burn, and a plant or two on very steep craggy slopes between there and The Priest (1994, LAI and WS). The Shetland plant is the attractive form with leaves blotched with purple.

Section Hieracium.

†**H. ronasii.** *Ronas Voe Hawkweed.* Endemic. Very rare. Rocky outcrops and crags (usually dry but occasionally damp), and (formerly and very rarely) on dry rocky streambanks. This scarce shy-flowering species has been recorded from both sides of Ronas Voe, Northmavine, and once from Unst. It was first collected, in 1865, by R. Tate, on 'granite cliffs, Ronas Voe', almost certainly from the N side of the *voe*. From then until 1979 it has been collected from this area on only a handful of occasions. During the period 1992–1994 L. A. Inkster and WS surveyed (usually jointly) this large craggy district and were rewarded by finding *H. ronasii* in nine spots or larger areas. These extended from the SE end of the Ayre of Teogs NW to the gully at the N end of the beach at Slocka, with most sites within 40–125 m altitude. In most locations a handful of rosettes only were found, but larger numbers grew at the following two of the nine sites: about twenty plants (only four or five in flower) on the very large and steep rock-face 125 m SE of the Burn of Teogs (1994, LAI and WS); and about forty plants (all rosetttes) in a crevice among rocks between the Y-shaped burn below The Brough and the burn immediately to the S, altitude about fifty m (1994, WS). In 1958 a few rosettes were found on the dry banks of a stream on the '29' vertical Ordnance Survey line at Heylor, by the S side of Ronas Voe; in 1992 the plant could not be found. In 1884 it was collected at 'Booner's Hill', Unst, by W. E. and H. Smith; neither the plant nor the locality has been heard of again. *H. ronasii* is the first of our hawkweeds to flower, both in the wild and in cultivation.

Antennaria dioica. *Mountain Everlasting.* Stony or rocky heaths, often mossy and hummocky and always on thin dry soils. It is frequent to common in parts of Northmavine (especially on Ronas Hill and the extensive plateau to the N), Fetlar, and Unst. It is found about Grimister, Yell, and is seen locally elsewhere in Shetland except where conditions are too peaty. Also in Fair Isle, Vaila, etc.

*****Anaphalis margaritacea.** *Pearly Everlasting.* Casual. An occasional garden plant which is sometimes seen outside gardens as a straggler or deliberate introduction.

Gnaphalium sylvaticum. *Heath Cudweed*. Nowadays this appears to be a very rare casual. There is a single recent record, the first since 1986: many plants on the car park at the Muckle Roe cemetery (2006, A. K. Thorne). By 2008, however, WS could not find a single specimen. In the past—it was recorded as early as 1769—heath cudweed appears to have been a native, but perhaps an alien in some cases, and was fairly evenly distributed throughout Mainland, Fetlar, Yell, and Unst. It grew in a number of habitats: dry, often rocky pastures and turfy slopes; also (but much more rarely) on sand-dunes and in arable and fallow ground, etc. In many places it occurred sparingly and only once, in 1961 at Arisdale, in the S of Yell, was it noted in abundance. At three places it had a long and presumably continuous existence: Tingwall, Central Mainland (1865, 1888, 1921); Burrafirth, Unst (1865, 1960, 1963), and at Sandwick, near Swining, North Mainland (1963, 1983). Many new stations were discovered in the 1960s and 1970s as a result of sustained botanical recording, the first serious survey for forty years. However, it then suffered an unexpected and sudden decline. In the 1980s only four sites were recorded, the latest of these (in 1986) being at the mouth of the Dale Burn, near Sandsound, Central Mainland; this was to be the last record for twenty years. Overgrazing (at that time) by an increasing number of sheep seems to have led to its decrease, coupled with the fact that it is a short-lived perennial which requires regular seed production in order to survive in any one station.

G. uliginosum. *Marsh Cudweed*. A well-established alien of bare muddy or gravelly places, especially roadside verges and rutted or unmetalled tracks; also in damp arable ground (especially cornfields), a habitat now rarely seen. In 1994 WS saw scattered plants in a damp depression at the N end of Little Holm, off Scatness, South Mainland. Occasional throughout the county, and recorded from Fair Isle, Foula, and Mousa.

Helichrysum bellidioides. *New Zealand Everlastingflower*. A small mat-forming evergreen shrub which (since 1975 at least) has been a well-established alien on the rocky or grassy banks of a stream quite near the main road W of Tagon, Voe, North Mainland, probably a garden outcast. However, in 2009 L. A. Inkster and WS noted that it had deteriorated and was much less in evidence, perhaps as a result of flooding.

Inula helenium. *Elecampane*. Casual. This is a relic or outcast of gardens, perhaps sometimes deliberately introduced. It is occasionally seen about crofts, in old yards, and by roadsides and field-borders, etc., and can persist for a very long time.

Solidago virgaurea. *Goldenrod*. A frequent plant of dry places: heathy pastures; rocky outcrops; rocky or heathery sides of burns; low grassy or heathery sea-banks; and holms in lochs. Not recorded from Fair Isle or Foula. Accompanying species often include *Jasione montana, Luzula sylvatica,* and *Deschampsia flexuosa.*

S. gigantea subsp. **serotina**. *Early Goldenrod*. Casual. Once recorded: grassy bank by seashore, E side of the Voe of Leiraness, Bressay (1983), as a garden outcast.

Aster novi-belgii. *Confused Michaelmas-daisy*. Casual. This commonly grown late-flowering garden plant is occasionally seen around crofts and houses (sometimes as a garden relic or deliberate introduction), on roadsides and by burns, and as an often long-persistent outcast or straggler. One of its best sites must be the Burn of Laxdale, Cunningsburgh, South Mainland, where there are numerous occurrences along the banks of this lowland stream.

*A. lanceolatus. *Narrow-leaved Michaelmas-daisy*. This very rare casual was once recorded (in 1967) as a garden escape or relic on an old wall by Burns Lane, Lerwick.
A. tripolium. *Sea Aster.* Very rare. One of our more special coastal plants. Steep sea-cliffs where other vegetation is largely absent and where the plant itself is mostly inaccessible. It was found on Isbister Holm, E of Whalsay in 1968, and in the following year on the E coast of Whalsay, opposite Isbister Holm. On the holm there could be as many as 5,000 plants on the side of a long narrow *geo* in the SE of the island. This site is best observed by entering the *geo* in a small boat on a very calm day, which also allows the lowest part of the colony to be safely and easily examined. On the W side of the holm, at the S end, and on much lower sea-cliffs, about 450 plants have been counted. In Whalsay about 500 plants grow in the Taing Geo/Yoxie Geo area.
Bellis perennis. *Daisy.* A common and often very abundant plant of short damp grassland; stony flushes; grassy places among dry rocky outcrops where it becomes very stunted during periods of drought; and on waste rubbly ground, roadside verges, and in lawns.
*Tanacetum parthenium. *Feverfew.* Casual. Occasionally seen as a mere garden outcast on rubbish-dumps and rubbly ground, and as an escape by garden walls in the Scalloway and Lerwick areas.
*T. vulgare. *Tansy.* A well-established alien in the N of Yell where it has been known, probably continuously, for well over a century. In 1997 R. C. Palmer found it thriving in a sandy valley near Breakon. There may be other places where it has a similar status, but in general it is best treated as an often long-persistent casual. It is a frequent garden straggler or outcast (perhaps sometimes a deliberate introduction), and is found not far from houses and crofts (often by stone walls), by roadsides, and on waste ground, etc. Occasional throughout the county.
*Artemisia vulgaris. *Mugwort.* A well-established but decreasing alien of waste grassy places about houses, by roadsides, and in or by cultivated ground, etc. Occasional, but absent or unrecorded from West Mainland. Mugwort has a marked preference for dry sandy soils, as about Grutness, South Mainland, in the N of Yell where, however, it is much scarcer now than in the past—in 1961 it grew plentifully in the Breakon area. It can also be found near 'The Haa', Wick of Skaw, Unst.
*A. abrotanum. *Southernwood.* Casual. A very rare garden outcast with two old records: Lerwick rubbish-dump, by the Loch of Clickimin (1955, 1958); on a grassy cliff below a garden, Walls, West Mainland (1973).
Achillea ptarmica. *Sneezewort.* A frequent plant in the county and recorded from all the larger islands. Damp grassy places (meadows, sides of burns and lochs) and in rough herbage by roadsides and field-borders.
A. millefolium. *Yarrow.* A common species of dry grassy places, notably near the coast and on limestone or sandy soils, and among rough herbage. Often very dwarfed.
*Anthemis arvensis. *Corn Chamomile.* A very rare casual. Garden weed, Virkie, South Mainland (1970); enclosure of the new Scalloway reservoir, a few plants (1961); and a bushy clump by the Unst Leisure Centre (1989, R. C. Palmer).
*A. cotula. *Stinking Chamomile.* A very rare casual with only one definite record: two or three plants on the grassy roadside verge by the Weisdale Church of Scotland manse, Central Mainland (2003, WS).

*Chrysanthemum segetum. *Corn Marigold.* A rare casual. Cultivated and waste ground, etc. Judging by the notes of those recorders who provided frequency details, this seems always to have been a scarce plant. It has probably not been more than a casual (which it certainly seems to be now) or at most a weakly-established alien. Its chief habitat was arable ground, especially cornfields, but the steady reduction in oat cultivation, coupled with cleaner seed, has almost removed corn marigold from the local scene. It was last recorded from cornfields in 1983 at Cullivoe, Yell, and Tow, Cunningsburgh, South Mainland. There are three post-1987 records: two plants by midden, Quoy, S end of Fair Isle (1997, 1998, N. Riddiford); one plant in a barley-field, North Exnaboe, South Mainland (1999, 2000); two plants by retaining wall by the road near the head of the tiny *voe* at the S end of Bruray, Out Skerries (2005, J. S. Blackadder).

*Leucanthemum vulgare. *Oxeye Daisy.* Casual. Thinly scattered over much of the county. Regardless of its long persistence in some areas, and its probably temporary abundance in others, WS does not regard this as a good example of a well-established alien. Dry grassland: meadows, pastures, churchyards, and on reseeded roadside embankments, etc.; sometimes abundant, but just as often in small patches or single plants, a situation which seems to have obtained for over a century.

*L. × superbum. *Shasta Daisy.* Casual. This is frequently grown in gardens and is sometimes seen as an outcast or deliberate introduction. It is capable of persisting for a long time as at Walls, West Mainland, where, as a blatant outcast, it grew from at least 1968 to 1990 on a low sea-bank near the pier. Not seen in 2008 (WS).

*Matricaria recutita. *Scented Mayweed.* This is a very rare casual which has been recorded on four occasions: two plants on a weedy roadside opposite Wishart's Stores, Cunningsburgh, South Mainland (1978); several plants on reseeded banks by the main road between the school and Stebbigrind, Whiteness, Central Mainland (1988, WS); a weed in the garden of 9 Knab Road, Lerwick (1998, I. Clark); among cabbages in the Fair Isle Bird observatory garden (1999, N. Riddiford).

*M. discoidea. *Pineappleweed.* A well-established and common alien of crofting and farming areas (especially on trodden tracks and trampled gateways, and by byres and barns), and as an arable weed; also by roadsides, on open waste ground, about harbours, etc. Evenly distributed throughout the whole of Shetland.

Tripleurospermum maritimum. Sell & Murrell (2006) recognise four subspecies in this country, two of which (subsp. *nigriceps* and subsp. *inodorum*) occur in Shetland. Subsp. **nigriceps** (sea mayweed), a plant of northernmost mainland Scotland, Orkney and Shetland, and which has for a long time been incorrectly equated with the Arctic subsp. *phaeocephalum*, is common on shingle beaches, sea-cliffs, large skerries and stacks, and is often abundant and robust when these are enriched with guano. A very similar, if not identical, plant grows in arable ground by the coast, a juxtaposition of frequent occurrence in the county. *Subsp. **inodorum** (scentless mayweed), here regarded as a casual, has been noted on a few occasions in arable ground in Mainland and in Yell, but its frequency and distribution are unclear through confusion with the supposedly agrestal form of subsp. *nigriceps*.

*Cotula squalida. *Leptinella.* A well-established alien occurring in a few places in Central and North Mainland, and in Northmavine, mainly on turfy roadside verges, often extending into and competing with natural vegetation and becoming completely naturalised. At the Setter Scord and Dales Voe stations it has increased markedly in recent years. In Central Mainland it is found near a house at the head of Cat Firth, South Nesting; on both sides of the B9071 at the Setter Scord, E of the Loch of Gonfirth (2001, WS), and by the same road leading down to Voe from the A970, but now less common in the last-named station than in the past. There are also three sites in North Mainland: by the A968 above Southlee and Rugg, Dales Voe, abundant; by the A968 where it crosses the Burn of Grunnavoe, above Tofts Voe; junction of the B9076 with a track, leading to Sella Ness, by the SW side of Garths Voe, covering several hundreds of square metres. It was first noted at Garths Voe in 1957, the first record from Shetland, and had already been established there for some time. In Northmavine it was found by the access road at the Valladale housing scheme, Urafirth (2006, WS). This plant seems set to remain in Shetland for a long time.

*Senecio fluviatilis. *Broad-leaved Ragwort*. A very rare casual of garden origin. Brettabister, North Nesting, Central Mainland, several large patches near a house and in nearby ditches, garden outcast (1971), still one large patch in 1997, but gone by (2008, WS). Since 1979 it has been a garden relic at nearby Kirkabister.

*S. smithii. *Magellan Ragwort*. A garden escape, relic of cultivation, or deliberate introduction in the wild, and now—and indeed for a long time—a well-established alien around houses, especially ruined crofthouses, by lochs, burns, and roadsides. Scattered throughout Shetland, notably in South Mainland, parts of West Mainland, and Yell. It can be very persistent, as by a roadside in Fair Isle since at least the 1930s and, from the same period, on the small holm in the Loch of Vatsetter, SE of Mid Yell. Scott & Palmer (1987) claim that it is likely to have arrived in Shetland through 'normal gardening channels', and not, as has been supposed, by locals returning from sheep-farming or whaling in the South Atlantic. It is a native species of the southern parts of Chile and Argentina.

*S. jacobaea. *Common Ragwort*. A well-established alien on or near steep grassy banks in the W part of Scalloway where it was first recorded, in 1888, by W. H. Beeby. Elsewhere in Shetland, including other places in and around Scalloway, it is a scarce casual by roadsides, in dry grassy places, by field-borders, and on rubbly ground around new buildings (notably in the N part of Lerwick), and about quarries, etc. It is often seen as single plants or as small groups or diffuse patches; rarely in significant abundance. There are a few records from Mainland (except the W part) and from Unst, but it is so far unrecorded from Yell.

S. × ostenfeldii (*S. jacobaea × S. aquaticus*). This appears to have had a continuous existence at Scalloway since at least 1888 when it was collected by W. H. Beeby. The subsequent published record by G. C. Druce was the first for Britain. It grows, usually sparingly, both within and on the outskirts of the village, by roads and field-borders, etc. Although normally persistent, it can appear in new localities and disappear from old stations from time to time.

S. aquaticus. *Marsh Ragwort*. A widely distributed and often common species of damp or dry, often rough, meadows and pastures (particularly in crofting areas), sandy links, stony loch margins, holms in lochs, etc.

*S. squalidus. *Oxford Ragwort*. A very rare casual with two records only, both from Lerwick: waste ground, Grantfield, rare (1968); by the main road near the Holmsgarth Ferry Terminal, one clump (1986–1987).

*S. vulgaris. *Groundsel*. A common and well-established alien of cultivated ground (arable fields and gardens) and open waste ground generally. Recorded from nearly all the inhabited and crofting areas.

*S. vernalis. *Eastern Groundsel*. A very rare casual. One record: half-a-dozen plants, Unst Leisure Centre (1989). Probably a grass-seed impurity.

*Tussilago farfara. *Colt's-foot*. Frequent in South Mainland, occasional elsewhere throughout the county. A well-established alien of gravelly places such as weedy roadside verges and banks, by burns, in grassy places around houses, and on waste ground and heaps of rubble, etc. It is sometimes seen a long way from houses.

*Petasites albus. *White Butterbur*. A long-persistent and well-established alien of garden origin. Waste grassy places and roadside banks in a very few locations, as about Scatness, Boddam, and between Setter and Leebotten (Sandwick), all in South Mainland, and in one or two spots in Unst. It was once a feature of the Stove area, Sandwick, South Mainland, but extensive roadworks appear to have led to its disappearance. There are two old records from Yell: Easterlee (1961), and Littlester (1962), both near Burravoe, and perhaps casual in each place. In 1956 it occurred as a casual on the Lerwick rubbish-dump (by the Loch of Clickimin).

*P. fragrans. *Winter Heliotrope*. Another long-persistent and well-established alien of garden origin. As such it is known from one place; the yard behind the 'Aald Haa', Scalloway, where it was first noted in 1963 and where it was already common. In 2009 it covered 350 square m; earlier, in 1984, it had occupied a little less than half of that area. Four old records exist: Lerwick, on a sea-bank (1898–9); Weisdale, Central Mainland (1924); Laxfirth, Tingwall, Central Mainland, of garden origin (1966); and Westshore, Scalloway, by the shore (pre-1958 to about 1965). Probably a casual in each of these four places. In February 2010 it flowered very freely at the 'Aald Haa'.

*Calendula officinalis. *Pot Marigold*. Casual. Cultivated in many gardens, but only rarely seen in their vicinity (unless deliberately introduced), or as an outcast on rubbish-dumps and waste rubbly places, etc.

*Ambrosia artemisiifolia. *Ragweed*. A bird-seed casual. Very sparingly near new roadworks, 'South Taing', Cott, Weisdale, Central Mainland (2009, R. Leask).

*Helianthus annuus. *Sunflower*. A very rare casual of garden origin. On the beach at the Sand of Meal, West Burra (1995); on the beach at Tarland, near Sound, Lerwick (1995); and in a field on the E side of the Houster road, Tingwall, Central Mainland, near its junction with the (old) A970 (2006, N. Anderson). Single plants at all sites.

*Galinsoga parviflora. *Gallant-soldier*. This very rare casual has been recorded once: garden weed, Eastshore, by the Pool of Virkie, South Mainland (1971).

*Aponogeton distachyos. *Cape-pondweed*. Casual. Planted [shortly after 1998] in the reservoir at Vaadal, near the Fair Isle Water Pumping Station (Tyler 2003).

Triglochin palustre. *Marsh Arrowgrass*. Frequent and widely distributed in wet places: marshes; meadows; wet stony heaths; saltmarshy turf; ditches, etc.

T. maritimum. *Sea Arrowgrass*. This is a frequent coastal plant of saltmarshes and saltmarshy turf by brackish pools and ditches; also, but rarely, some way inland as in marshy ground by the W side of the Loch of Tingwall, and on damp stony ground on the W side of the East Hill of Voe (S of Voe), both sites in Central Mainland.

Potamogeton natans. *Broad-leaved Pondweed*. Common throughout Shetland in lochs and pools; in Stanevatstoe Loch, near Sandness, West Mainland, it is abundant.

In the more peaty and swampy habitats it is largely replaced by the even more widespread and common *P. polygonifolius.*

P. × gessnacensis (*P. natans × P. polygonifolius*). Extremely rare. It was found in the Loch of Gards, at Scatness, South Mainland, in 1996 by P. M. Hollingsworth and C. D. Preston. At that time it had been recorded from only two other sites in Britain.

[**P. × sparganiifolius** (*P. natans × P. gramineus*). Believed to be extinct. An interesting pondweed hybrid, believed to be of this parentage, was found in the Loch of Cliff, Unst, in 1974 by R. H. Britton. It grew off the pier at the end of the track from Baliasta, at the S end of the loch, and was flourishing and flowering in 1976 when R. C. Palmer visited the site, although only about six plants were noted. He saw it there again two years later, but in 1981 and 1989 he searched for it without success.]

P. polygonifolius. *Bog Pondweed*. A very common species of peaty places: pools, lochs, ditches, and (occasionally) in running water; also commonly as a subterrestrial state in marshy or boggy places (especially where bog-iron oozes).

P. × zizii (*P. lucens × P. gramineus*). *Long-leaved Pondweed*. Very rare. Known only from the highly calcareous Loch of Hillwell, South Mainland. It was first noted here in 1962; in 1977 it was stated to be in fair quantity on the S side of the loch, and by the 1990s there had been a substantial increase. By 2002 it was dominant over much of the S half of the loch, and in 2006 N. Aspey reported 'lots of material washed up around the shore'. Its seemingly recent arrival and subsequent spread matches that of *Hippuris vulgaris* in the same loch, and leads to the possibility that *P. × zizii* may be found in other lochs in the area.

P. gramineus. *Various-leaved Pondweed*. A frequent plant of the richer lochs and streams. It largely avoids nutrient-poor as well as strongly peaty waters and is thus scarce or unrecorded in parts of Northmavine and Yell.

P. × nitens (*P. gramineus × P. perfoliatus*). *Bright-leaved Pondweed*. This hybrid grows in the same general habitat as that of its parents. There are twenty reliably recorded stations. Half of these are in Central and West Mainland; two in South Mainland; three in Northmavine; two in Unst; and one each in Foula, Yell, and Fetlar.

P. alpinus. *Red Pondweed*. Very rare and seemingly becoming rarer. This was first detected in Shetland in 1966 when R. C. Palmer found it among emergent vegetation in the NW corner of the Loch of Voe, Central Mainland. In 1967 he noted it in the first 100 m or so of the adjacent outflowing burn running along the E side of the A970 at this point. The same recorder, in 1976, observed two or three square m of it in the NW corner of the loch; in 1981 in the same place he said it was 'forming a dense bed and flowering freely', a situation also noted by WS during the following few years. There then followed a period with no observations. In 2000 WS saw a few patches in the stream, as did N. Aspey four years later; in 2002 P. V. Harvey and S. Whild reported two or three specimens in the corner. Perhaps the early 1980s was a favourable period for this rarity.

P. praelongus. *Long-stalked Pondweed*. A local pondweed of fifteen mainly large and moderately nutrient-rich lochs. There are two sites in South Mainland; two in Central Mainland; seven in West Mainland; one in Northmavine, and three in Unst. Two of these sites have been discovered fairly recently: Punds Water, E of Hamar, Northmavine (1997, Scottish Natural Heritage loch survey); and Sand Water, N of Hulma Water, West Mainland, many patches (2003, WS).

P. perfoliatus. *Perfoliate Pondweed.* Common. Widely distributed in moderately nutrient-rich lochs; rarely in large deep burns. Absent from some smaller islands.

P. friesii. *Flat-stalked Pondweed.* A very rare plant of moderately or very rich lochs. First recorded for Shetland in 1975 from the Loch of Clickimin, Lerwick, from where there are no later records. In 1977 it was noted in the Loch of Hillwell, South Mainland, apparently in small quantity, and perhaps a recent arrival. In 1982 it was abundant, as it was also in 1997 (Scottish Natural Heritage loch survey), and in 2004 there was 'quite a quantity in the strandline at the S end of the loch' (P. V. Harvey and S. Whild). It is probable that the severe eutrophication of the loch in the 1970s, now much minimised, may have been favourable to this plant. In 2004 N. F. Stewart recorded it as occasional to widespread in the Loch of Spiggie, South Mainland. It seems almost certain that this is the same plant which P. M. Hollingsworth and C. D. Preston saw in 1996 in the Loch of Spiggie, at a point SE of Symblisetter. The latter remarked that it fell between *P. friesii* and *P. pusillus*. N. F. Stewart, also in 2004, found *P. friesii* at the S end of the Loch of Tingwall, Central Mainland.

P. rutilus. *Shetland Pondweed.* A very rare species of two moderately rich lochs in Central Mainland (Lochs of Tingwall and Asta) and two very rich lochs in the Walls area of West Mainland (Lochs of Kirkigarth and Bardister). It is recorded as being frequent in all four lochs; in 1996 P. M. Hollingsworth and C. D. Preston found it in abundance by the S side of the Loch of Kirkigarth. In 1955 it grew in the burn from the Loch of Asta. In 1890 W. H. Beeby collected it the Loch of Bardister but was unaware that he had added a new species to Britain. In fact, he died before his material had been correctly identified. Elsewhere in Britain it grows in a few sites in the Inner and Outer Hebrides and mainland Scotland.

P. pusillus. *Lesser Pondweed.* A local pondweed of moderately rich to very rich, sometimes brackish, lochs and pools; not infrequently in shallow water overlying sand. Most of the records are from South and West Mainland, Yell, Fetlar, and Unst.

P. berchtoldii. *Small Pondweed.* Another local species, and found in fairly similar habitats to those of the preceding, sometimes in the same loch. Most records are from Central and West Mainland, Northmavine, and with occurrences in Whalsay, Yell, Unst, and several of the lesser islands. It has not been certainly recorded from the S of South Mainland where *P. pusillus* has a strong presence.

P. filiformis. *Slender-leaved Pondweed.* An occasional but widely distributed plant of moderately rich to very rich, sometimes brackish, lochs and pools, and slow-flowing burns and ditches; not uncommonly in shallow water overlying a sandy or muddy substrate.

P. × suecicus (*P. filiformis* × *P. pectinatus*). *Swedish Pondweed.* A very rare hybrid which was first collected in 1980 by R. C. Palmer by the S side of the Loch of Clickimin, Lerwick. However, it was not until 1996 that the true identity of his material was realised. In that year it was collected in the same loch by P. M. Hollingsworth and C. D. Preston. Their plants came from two sites in the N of the loch, one being the westernmost bay behind the causeway at the N end.

P. pectinatus. *Fennel Pondweed.* Rare. Habitats similar to those of *P. filiformis*. Recorded from the Lochs of Spiggie and Hillwell (South Mainland); the Lochs of Asta, Tingwall, Clickimin, and a coastal pool at the Ness of Sound, Lerwick (all Central Mainland); Croo Water, Funzie Ness, and the Loch of Funzie (both in Fetlar), the latter record made in 1996 by P. M. Hollingsworth and C. D. Preston; and the Lochs of Watlee and Snarravoe (Unst, 1997, Scottish Natural Heritage loch survey).

Ruppia maritima. *Beaked Tasselweed*. A local coastal plant which inhabits small lochs and pools behind beaches, in brackish or muddy, often stagnant, water; saltmarsh pools and channels; and (rarely) sheltered embayments of large tidal lochs. Although most records are from Central, West, and North Mainland, it also occurs in Whalsay, Fetlar, Yell, and Unst.

R. cirrhosa. *Spiral Tasselweed*. Occasional and with a more restricted distribution than that of *R. maritima*. It grows in the same general habitats but tolerates a more saline environment; it is thus often accompanied by seaweed, and is less often seen in lochs and pools behind beaches. There are twelve sites on record: two in Central Mainland; five in West Mainland; one in North Mainland; and four in Northmavine. Two of the best sites are at the head of Whiteness Voe, Central Mainland (where it grows with *R. maritima*) and in the Marlee Loch, West Mainland. Included in the foregoing is one new record: Effirth Voe, West Mainland (2006, BSBI field meeting).

Zannichellia palustris. *Horned Pondweed*. An occasional, but sometimes locally abundant plant of nutrient-rich or brackish lochs, very rarely in small pools. Listed from South, Central, and West Mainland, and from Yell, Unst and the small island of Haaf Gruney. It used to grow in the Loch of Urie, Fetlar. Four new records are included in the foregoing: Easter Loch, Uyeasound, and the Loch of Snarravoe (both in Unst, 1997, Scottish Natural Heritage loch survey); one plant on the strandline, Loch of Hillwell, South Mainland (2004, P. V. Harvey and S. Whild); and in Kirk Loch, Yell (2004, N. F. Stewart).

Zostera marina. *Eelgrass*. Very rare. This is our only flowering plant to grow in salt water and is very rarely exposed (at least nowadays), even at very low tides. It grows on a muddy-sandy substrate near or at the head of four large and relatively sheltered *voes*, and two sea-lochs. There are comparatively recent records from South Voe, between East and West Burra, in several places; Whiteness Voe, Weisdale Voe, and the associated Loch of Hellister (1993, C. Howson), these three sites in Central Mainland; S side of Effirth Voe, Bixter; Marlee Loch, at head of the Voe of Brindister, both in West Mainland. Three much older records exist: Tresta Voe, SE of Bixter, Central Mainland, abundant (*c*.1912); W side of The Firth, below Semblister, West Mainland (1927); and at Balta Sound, Unst (1887). No later records are known from these areas. In the early 1930s a disease severely reduced the colonies at Weisdale Voe and at the Marlee Loch. In fact, the latter site takes its name from 'marlie', the local name for eelgrass, an abundant plant there at one time. Fragments have been washed up on the North Haven beach, Fair Isle, and on the St Ninian's Isle tombolo in South Mainland. These are unlikely to have had a local origin, and probably came from Orkney where the plant is more widespread than in Shetland, and where the wide bays (as opposed to our narrow *voes*) would allow fragments to have easy access to the open sea.

Lemna minor. *Lesser Duckweed*. Very rare and a probable native except in the one site where it is of casual occurrence. It was first discovered, in 1955, in the Loch of Clickimin, Lerwick, behind the causeway at the N end, in very small quantity. The duckweed was there in 1984 (A. Douse) but still in tiny amounts. In some of the intervening years at least, and in 2008 (WS), it could not be found. In 1960 it was abundant in a ditch between the Loch of Clickimin and the North Loch, but the site was lost when the area later became part of the Clickimin Leisure Complex. In Bressay in 1963 it was found to be common in a pool known as the 'Dutchmens' Quarry', near 'Gardie House'; it was still there in 2000 (WS). Carrick *et al.* (1997) records it from the 'Tingwall Valley', Central Mainland. The site is just behind the shore of Lax Firth, a little way S of The Lotrans; here it is a casual which originated as a throw-out from a nearby garden *c.*1995 and which was still present in 2008 (WS). The record from Flaga Water, near West Sandwick, Yell (1997, Scottish Natural Heritage loch survey) remains unconfirmed. Lesser duckweed was listed from Shetland (without locality) in 1769 by J. Robertson, but WS is inclined to regard his report with some doubt.

Juncus squarrosus. *Heath Rush.* An abundant plant of peaty acid soils throughout the county: hill grassland (often with *Nardus stricta*); moist or wet heaths and moors; and boggy places among the hills.

J. **gerardii**. *Saltmarsh Rush.* Frequent. As the name implies, this is a species of the shoreline: short turf; saltmarshes and saltmarshy turf; muddy seashores; and among rocks. Often drenched in spray in autumn gales.

J. **trifidus**. *Three-leaved Rush.* Very rare, but occasional to frequent over a very large area of bare exposed granite fellfield in its only station, the summits and slopes above 240 m on Ronas Hill, Mid Field, and Roga Field, all parts of the Ronas Hill group, Northmavine. It extends from Abram's Ward, on the SW shoulder of Ronas Hill, NE to Roga Field. *J. trifidus* is one of our very few arctic-alpines which does not descend to near sea-level.

J. **bufonius**. *Toad Rush.* Common over a range of damp, wet or muddy places: open waste or disturbed ground (especially where much frequented by sheep); saltmarshes and spray-washed turf; stony or muddy beaches; by pools and ditches; and in arable ground, etc.

J. **ambiguus**. *Frog Rush.* Rare on present knowledge, but probably overlooked by being mistaken for the similar and much more common *J. bufonius* which, moreover, can also grow in the same habitat. Apparently restricted to the coast where it occurs by the muddy edges of small pools and lochs and in saltmarshy turf by the mouth of streams. It is recorded from eight stations: Loch of Gards, Scatness (1996, P. M. Hollingsworth and C. D. Preston), pool at Grutness (2006, P. V. Harvey and S. Whild), both in South Mainland; behind Burgi Ayre, Mousa (2002, R. Norde); Hildasay (1888); below Tronafirth, Lax Firth, Central Mainland (1981); Symbister, Whalsay (1986); mouth of the Burn of Laxobigging, Graven, North Mainland (1980); and at Uyeasound, Unst (1894). This species is likely to be more frequent than the foregoing records would suggest.

J. **articulatus**. *Jointed Rush.* An abundant and widespread plant of watery habitats; marshy or swampy ground, sides of pools and burns, ditches, wet cliffy sea-banks, and as a submerged form in lochs and burns. It tends to avoid the most peaty situations.

*****J**. **acutiflorus**. *Sharp-flowered Rush.* A local and well-established alien of damp moory or turfy pastures and wet meadows, almost always by or near roads and invariably in or close to inhabited areas. When undisturbed it forms large and long-persistent patches. It was first recorded, in 1957, not far from the (then) Lerwick Geophysical Observatory, and is still there to this day as a large and obvious feature.

80

There are twenty-one records spread over the county from South Mainland to Unst. On numerous occasions over a long period it has been misnamed *J. articulatus* despite the obvious dissimilarities between them. In 2009 WS confirmed its presence in twelve of its recorded sites, but could not find it in a further five where it may have been removed by roadworks. It was not looked for in the remaining four sites. This rush appears to have been a recurring accidental introduction over three or four decades. The two latest records are from Baltasound, Unst (1992, WS) and from above 'The Lodge', Windhouse, Yell (1998, R. C. Palmer). Assuming that its origin lay outwith Shetland, its period of expansion may be coming to an end. Although all of its records have been placed under the present species, there is a possibility that the hybrid between it and *J. articulatus* (*J. × surrejanus*) may be present in Shetland. However, this hybrid, which grows in Orkney, has not yet been definitely confirmed from our area. The writer is indebted to M. Wilcox and C. Stace for much help with *J. acutiflorus* in general.

J. **bulbosus**. *Bulbous Rush.* An abundant and highly variable plant of wet, often peaty, habitats. As a terrestrial species it is found in boggy places on heaths and moors, on rutted tracks, and by the stony edges of lochs, etc. As an aquatic it occurs as a long, straggly form in burns; the same form is found in lochs and pools where it is also frequently seen as detached, free-floating, but living, fragments. An extreme state is seen on the muddy bottom of lochs and pools—tufted, hair-like, barren. and totally submerged. These forms are so unlike each other that the differences between them are scarcely credible.

J. **triglumis**. *Three-flowered Rush.* A very rare arctic-alpine species known from only one locality, in damp stony flushes on metagabbro on the NNE slopes of the Hill of Colvadale, Unst. It was first found, in 1887, by W. H. Beeby, possibly seen by D. H. N. Spence in the early 1950s, and certainly collected in 1962 by R. C. Palmer. In 2003 C. Geddes and A. Payne discovered three small groups of plants totalling at least 150 flowering clumps; two inside the E edge of HU6106, the third in the SW corner of HU6206. Later in the same year L. A. Inkster and WS located the three groups and found another (small) group near the first two.

J. **effusus**. *Soft-rush.* Common to abundant on damp heathy pastures and wet moorland, and particularly characteristic of the sides of peaty burns where it often forms large stands. Its habit varies from erect to widely spreading, and in the latter form the stems are often very gently spiralled or curved.

J. × **kern-reichgeltii** (*J. effusus* × *J. conglomeratus*). This not unexpected hybrid was first found in Shetland in 2009 by the writer. At least twelve clumps, some quite large, grew on damp moorland NE of Shalder's Ayre, W of Scalloway (HU39243972). A wide-ranging systematic search would almost certainly reveal further stations in the county. Plants with dull (not glossy) stems which are clearly ridged should be closely examined; the exact nature of the ridging should indicate whether they belong to *J. conglomeratus* or to the present hybrid. The writer is much indebted to M. Wilcox for identifying the Scalloway plant, and for his helpful comments.

J. **conglomeratus**. *Compact Rush.* Frequent and almost certainly widely overlooked. While its distribution and habitats are broadly similar to those of *J. effusus,* the present species seems to prefer drier ground and, unlike soft-rush, does not appear to form extensive populations.

Luzula pilosa. *Hairy Wood-rush*. Occasional to frequent. Among short heather on dry heaths and moors, especially in parts of West Mainland. It also grows in Central and North Mainland, but is unrecorded from South Mainland, Fetlar, and Unst.

L. sylvatica. *Great Wood-rush*. Common and widespread. Rocky places (notably steep-sided burns, sheltered ledges on crags, and on sea-banks and sea-cliffs); holms in lochs where it is often abundant if not the dominant species; and occasionally carpeting high peaty moorland slopes. There is evidence to suggest that when the vegetation on a holm is burned, any surviving great wood-rush is liable to colonise the bare ground and eventually become dominant.

L. campestris. *Field Wood-rush*. A common and widespread plant of dry grassland, particularly on the better soils, but also found in damper ground and on slightly acidic and heathy pastures.

L. multiflora. *Heath Wood-rush*. Another common and widespread wood-rush of dry or damp places. The typical form, subsp. **multiflora**, grows among heather in moorland areas. Subsp. **congesta** favours dry short heathy pastures in coastal areas; here it can be very dwarfed in exposed maritime turf. The two subspecies cannot always be satisfactorily separated in coastal situations where their habitats overlap.

L. spicata. *Spiked Wood-rush*. A very rare arctic-alpine species which seems to be confined to one large area, the bare exposed granite fellfield on the summit and slopes of Ronas Hill, Northmavine. It occurs from near the cairn on Abram's Ward, on the SW shoulder of the hill, NE to near the loch (the highest in Shetland) at the Shurgie Scord, between Ronas Hill and Mid Field. *L. spicata* is very sparingly scattered over this extensive area; even a diligent search may reveal a few plants only, or none. However, in 2003 P. V. Harvey and WS found about fifty plants and some barren tufts over a very small flat area of fellfield a little to the SW of the above-mentioned loch.

Eriophorum angustifolium. *Common Cottongrass*. Abundant and widespread in wet or boggy moorland, especially where the peat is deep and wet, and in the water by the margins of soft-edged peaty pools and lochs.

E. vaginatum. *Hare's-tail Cottongrass*. Common and widespread on damp moors and heaths, and prefers drier ground than that favoured by the preceding species.

Trichophorum germanicum. *Deergrass*. Common to abundant on damp heaths and moors throughout Shetland, and often surviving after heather has been burned.

T. cespitosum. *Northern Deergrass*. Apparently very rare on our present and extremely limited knowledge. Its habitats are presumed to be similar to those of *T. germanicum*, but it is likely to prefer less strongly peaty areas. Four records: Bressay (1953, M. A. E. Richards); SE of Swabie Water, North Roe, Northmavine (2003, P. V. Harvey, A. Lockton, and S. Whild); at HU417422 [above the SE side of the Loch of Tingwall], Central Mainland (2007, R. Walls); and above Shalder's Ayre, W of Scalloway (2008, WS), the last two in Central Mainland.

T. × foersteri (*T. cespitosum × T. germanicum*). The opening comments on the above taxon also apply here. There are just three records: Burn of Roerwater, North Roe, Northmavine, at HU351848 (2003, P. V. Harvey, A. Lockton, and S. Whild); above the SE side of the Loch of Tingwall, Central Mainland, with *T. cespitosum*; and near the upper end of the Burn of Fitch, E of Scalloway, both finds made in 2008 (WS).

Eleocharis palustris subsp. **vulgaris**. *Common Spike-rush*. This is a common and widespread species, sometimes seen in abundance. Is is found in low-lying swampy ground, often by lochs and pools; also in shallow water overlying mud or sand at the edges of (usually) nutrient-rich or even brackish lochs. Some coastal populations appear to be intermediate between this and the following taxon.

E. uniglumis. *Slender Spike-rush.* A rare plant of grassy saltmarshes and turf by brackish lochs. In West Mainland it is recorded from the head of Seli Voe, near Garderhouse, and from the Bight of Braewick, W side of Aith Voe. There is one record from North Mainland: at the head of Dales Voe. In Northmavine there are four records: sea-loch, Housetter; mouth of the burn below South Collafirth; Loch of Queyfirth; and at The Houb, S of Sullom. In 2000 WS could not find it at South Collafirth where its habitat had been turned into rich pasture. There are many other records for *E. uniglumis* but all are dubious or refer to the preceding species. The Cheynies record in Scott & Palmer (1987), although based on expertly examined material, may refer to the common species.

E. multicaulis. *Many-stalked Spike-rush.* A widespread and locally frequent species of peaty ground: boggy moors, by moorland burns, and the sphagnous edges of lochs and pools. It is particularly prevalent in parts of West Mainland, North Mainland, and Northmavine, but is unrecorded from Fetlar.

E. quinqueflora. *Few-flowered Spike-rush.* Frequent in base-rich marshy places: stony flushes (by lochs, among moors and on coastal slopes); peaty swamps; and not uncommon in saltmarshy turf. It is evenly distributed throughout the whole county.

E. acicularis. *Needle Spike-rush.* Very rare and easily overlooked. Known with certainty from two contiguous nutrient-rich lochs in South Mainland—the Lochs of Spiggie and Brow—and in both of which it occurs as submerged and terrestrial forms. In the barren aquatic state it is locally frequent to common on the sandy-muddy bottom by the S end of the Loch of Spiggie, and in the same habitat by the SW side of the Loch of Brow. As a usually flowering terrestrial plant it is very sparingly found on the sandy-stony margin of the Loch of Spiggie at Littleness and Symblisetter, and by the NE corner and the SE side of the Loch of Brow.

Bolboschoenus maritimus. *Sea Club-rush.* Very rare. Discovered in 1997 by WS in brackish conditions at the head of Ura Firth, Northmavine. In 2006 he recorded a large elongated patch fifty-five paces in circumference and six paces at its widest at the E end of The Wadill, where the Dale Burn enters this brackish loch. There were also three narrow and rather open strips in the sluggish burn running from The Wadill to the sea, about opposite a walled yard; one was eighteen paces long, the second twelve, and the third five paces long. In 2009 L. A. Inkster and WS noted that the position remained the same except that the third strip had increased to twelve paces in length. They also saw that the plant had begun to colonise the small embayment immediately N of the large elongated patch. This is a recent and natural extension into a safer area, and especially pleasing as the colonies by the sluggish burn were always at some risk of being cleared by drainage, as happened on one occasion. Its attempt to spread may indicate that sea club-rush is a relatively recent arrival in Shetland.

Schoenoplectus lacustris. *Common Club-rush.* Rare. Known from eight moderately rich lochs and one swamp. Five sites are in Central Mainland: Flossy Loch, SE of Scalloway; Tirsa Water, E of Stromfirth; Sand Water, between Weisdale and South Nesting (by far the best site in Shetland); Sae Water, near Voe; and the Loch of Kirkabister, North Nesting. Three are in West Mainland: Loch of Houster, Aith;

Grass Water (HU2853); and at the Smalla Waters, by Sulma Water. In Unst it occurs in the swampy valley, known as the Loch of Vinstrick, NNE of Lund, and where it was in small quantity in 2004 (WS). It used to grow in Trolla Water, near Girlsta, Central Mainland, where it occurred up to 1956, but has not been seen since.

Isolepis setacea. *Bristle Club-rush.* Rare on present evidence but likely to escape notice because of its small size and inconspicuousness. Boggy, but not strongly peaty ground (often by burns or near the sea), by or in ditches and flushes, and in damp gravelly places. In many of its sites it has not been seen for a long time. It is recorded from eleven places, all in Mainland. In South Mainland it has three stations (Boddam, North Voxter (Cunningsburgh), and between Easter Quarff and Wester Quarff). Seven locations are in Central Mainland. Four of these are in the South Nesting area, with an outlier at the Burn of Pettawater (between Petta Water and Sand Water). The remaining two localities in this division are in the Scalloway district. From West Mainland there is a single record, from between Walls and the Bridge of Walls.

Eleogiton fluitans. *Floating Club-rush.* Locally frequent in some areas, absent or unrecorded from others. A species of peaty places such as burns, lochs and pools, and with a distinct preference for sandstone districts. In South Mainland it grows about the Lochs of Spiggie and Brow, and is on record from Fair Isle and Trondra. The Clock Loch, near Gletness, South Nesting, and the Loch of Kirkabister, North Nesting, are the only known stations for Central Mainland. From Bressay it is known from near the Loch of Seligeo (1993, L. Farrell). In West Mainland it is locally frequent from Hulma Water and Grass Water, through Walls and to Stanevatstoe Loch. From North Mainland there is one record: Winneries, Lunna Ness, and one from Northmavine, at Roer Water (2004, N. F. Stewart). It is found in the Town Loch, Muckle Roe (1993, LF), and in Vaila.

Blysmus rufus. *Saltmarsh Flat-sedge.* A local plant of the coast. Muddy saltmarshy turf, often at the head of a *voe* or by a sheltered embayment near the mouth of a burn, rarely by tidal lochs. There is a total of xvideos.comeighteen records: South Mainland (1), Central Mainland (6), West Mainland (3), North Mainland (4), Northmavine (3), and Unst (1). Two of these records remain unconfirmed: Aith Voe, Cunningsburgh, South Mainland (1953), and Bressay (*c.*1808). In 1989 it was in very limited quantity in its only Unst station, at Baltasound (R. C. Palmer).

Schoenus nigricans. *Black Bog-rush.* Frequent and widely distributed. Wet stony flushes among the hills , near lochs, and on coastal slopes, preferring base-rich soils and thus avoiding strongly peaty situations; also in marshes and bogs, and particularly characteristic of wet or dry serpentinite heaths in Unst and Fetlar.

Carex paniculata. *Greater Tussock-sedge.* Very rare. Peaty swamps and marshes. Known only from Foula (marsh at Ristie); Fetlar (swamps by the Burn of Funzie and at Trona Mires, and (1991, M. Tickner) in the Loch of Stivla area); and Unst (swampy ground on the N side of the Burrafirth road near Haroldswick).

C. diandra. *Lesser Tussock-sedge.* Very rare. Frequent over a large marshy area by and near the E side of the Burn of Mailand, SW of the Mailand croft, Unst. Also W of the same burn, W of Baliasta, very scarce (1997, R. C. Palmer).

C. arenaria. *Sand Sedge.* Locally common in sandy coastal habitats and, not surprisingly, frequent in the S half of South Mainland, notably about Sumburgh and the Bay of Quendale. Sand sedge also grows at the Voe of Sound, Lerwick; Rea Wick, and Sandness, both in West Mainland; Papa Stour; West Sand Wick, Yell; Wick of Tresta, Fetlar; and from Burra Firth, Nor Wick, and the Wick of Skaw, all in Unst.

C. maritima. *Curved Sedge*. A very rare arctic-alpine species which is both a native and an alien in Shetland. As a **native** it occurs in damp hollows or by small streams and seepage channels among sand-dunes or other extensive areas of sandy coastal turf; rarely (as in Mousa) in shell-sandy turf by a rocky beach. Five of its recorded twelve sites are in the S portion of South Mainland, and three of these were its strongholds in Shetland: Sumburgh, Links of Quendale, and between the N end of the Loch of Spiggie and the Bay of Scousburgh. In these three areas its habitat has been nearly or entirely destroyed by agricultural or industrial operations; at Sumburgh it would have been present in the early 1970s but has not been seen since. In 2000 a single plant was seen at Spiggie, and at Quendale a few patches can still be found. In 1982 a fine colony grew near the large marsh SW of the Loch of Clevigarth, NNE of North Exnaboe; in 2005 the writer could not find the sedge but felt it could still be present. It is also on record from St Ninian's Isle, and fine colonies exist at the Sand of Meal (West Burra), and on Mousa. Elsewhere in Shetland it is recorded as follows. Kirk Sand, Papa Stour (thousands of plants, 2002, WS); Sand Wick, near Hillswick, Northmavine (1972, no later record); West Sand Wick, Yell, where in 2000 WS noticed between thirty and forty plants only, a large reduction in the size of the colony, either through wave action or by the burn in spate); Links of Tresta, Fetlar, a small amount; and Nor Wick, Unst (not seen since 1974). There are two early records. In 1769 it was seen 'near Skelberry', South Mainland, almost certainly by J. Robertson; his station would have been the one at the N end of the Loch of Spiggie. Somewhat later, in 1806, C. Fothergill found it at the Bay of Quendale, in the same division, and commented that it was 'frequent in many sandy spots throughout Shetland' but provided no further details.

Curved sedge is a well-established **alien** in two places: at the waterworks, Helliers Water, Unst, in muddy-gravelly ground near a concrete foundation, at least 500 very small plants (1992, WS, and first noted here, by R. C. Palmer, in 1989); about 100 plants on grassy-stony ground by the S side of the track leading to the summit of the Ward of Scousburgh, South Mainland, close to its junction with a track to Twarri Field (1998, D. Mathias, F. H. Perring, and T. Russell). In 1982 two specimens grew by a track in a field (with wartime remains) N of the Burn of Hillwell, South Mainland, but has not been recorded again from this third, and probably casual inland site. In each case the sedge has apparently been accidentally introduced from the coast during construction operations.

C. ovalis. *Oval Sedge*. Common and widespread in damp (even marshy) tussocky, often acid, meadows and pastures, especially in low-lying and poorly drained districts.

C. echinata. *Star Sedge*. Common and widespread. A sedge of damp or wet peaty habitats: heaths, moors, and swampy or boggy places.

C. dioica. *Dioecious Sedge*. A frequent sedge of damp or wet ground: stony flushes, seepage channels, marshy or boggy ground. It prefers base-rich areas but can grow among sphagnum.

C. curta. *White Sedge*. A rare sedge of somewhat nutrient-rich swamps, often with *C. rostrata* and *Potentilla palustris*. In South Mainland it has been found between the Lochs of Spiggie and Brow (2004, A. Acton). The Flossy Loch and the Lang Lochs,

SE of Scalloway, are the only known Central Mainland stations. There is one site in West Mainland: the large swamp at Culswick (1998, M. Adam and P. V. Harvey). In Northmavine it grows by the Loch of Houllsquey, North Roe. In Fetlar white sedge occurs by the NW side of Papil Water; at the source and by the upper reaches of the nearby Burn of Vaus; and near the track to The Dale, Lamb Hoga, three or four clumps (1997, R. C. Palmer). In Unst it is frequent to abundant in the Haroldswick and Norwick areas. In 1958 it was found in the Calback area, North Mainland, but has not been seen again and seems likely to have disappeared during the construction of the Sullom Voe Oil Terminal.

C. rostrata. *Bottle Sedge.* Peaty lochs, pools and burns (both by their boggy margins and in the water, often forming large stands); also in more nutrient-rich swampy or marshy ground. Frequent generally but absent from Fair Isle, Foula, Out Skerries, etc.

C. × involuta (*C. rostrata* × *C. vesicaria*). Plants assumed to be of this parentage have been found in a few places in habitats similar to those favoured by *C. rostrata*: near the Burn of Strand, Tingwall, Central Mainland; Sma Lochs (W of Lunga Water), Smalla Waters (SW of Unifirth), and Grass Water, all in West Mainland; Hogalee Burn, S of the Loch of Vatsetter, Yell (R. C. Palmer, 1994); and in the Burn of Mailand where it joins the Loch of Cliff, Unst. It would appear that *C. vesicaria* itself does not grow in Shetland.

C. flacca. *Glaucous Sedge.* A fairly common sedge of base-rich damp or wet places (flushes among rocks or on low sea-banks), also in drier situations (especially in limestone grassland, sandy pastures, and on the serpentinite heath in Unst), less often in mildly peaty habitats.

C. panicea. *Carnation Sedge.* Another fairly common sedge of a wide range of habitats from marshes and bogs to wet or dry stony heaths and damp moorland.

C. binervis. *Green-ribbed Sedge.* A frequent and widespread sedge of dry or moist places on heaths, moors, and peaty grassland; also commonly found on ungrazed heathery rock-ledges and rocky streamsides, and on ungrazed peaty holms in lochs. In low-lying sheep-free places (as on some holms in lochs) it can grow to 155 cm in length.

C. hostiana. *Tawny Sedge.* Frequent and reasonably widespread, but often in small quantity, in base-rich marshes and flushes, on damp grassy slopes, and (sometimes) in mildly peaty habitats.

C. hostiana × **C. viridula** subsp. **brachyrrhyncha**. Widespread in flushes, base-rich marshes, and in swampy or boggy places.

C. hostiana × **C. viridula** subsp. **oedocarpa**. Widespread in flushes, marshes, and in boggy or moory places.

C. viridula. *Yellow-sedge.* This consists of three widespread subspecies which are not always easily told apart. Subsp. **brachyrrhyncha** is frequent in flushes, base-rich marshy or swampy ground, and (very rarely) on boggy peat-moors. It is not uncommon on the Central Mainland limestone and is found on other basic rocks elsewhere. Subsp. **oedocarpa** is a common plant of damp moors and heaths, stony shores of peaty lochs, and (rarely) in damp coastal pasture where it is often replaced by the following. Subsp. **viridula** is frequent in cliff-top pasture and in wet stony ground near the sea, and (rarely) in saltmarshes; also occasionally inland on sandy-stony loch shores and in flushes and damp turf by their margins, rarely in peat-bogs.

C. pilulifera. *Pill Sedge*. Frequent and widespread on dry base-poor soils: acid pastures, grassy areas on heaths and moors, and on stony hill slopes and summits.

C. limosa. *Bog-sedge*. Occasional in West Mainland, very rare elsewhere. Peaty swamps; also in bog-pools and by their swampy margins. In South Mainland it grows between the Lochs of Spiggie and Brow, and was recently discovered (2006, WS) in the extensive swamp at Vaamir, ENE of South Scousburgh. There is one station in Central Mainland: by the Burn of Pettawater, between Sand Water and Petta Water. In West Mainland (where eight of its seventeen sites are located) it is recorded as follows: near Lera Voe, WSW of Walls; between Mousavord Loch and the road, and between Burga Water and the road; pool E of the N end of the Loch of Voxterby (2003, WS); by roadside near Longa Water, West Burrafirth; lochan N of the Dogg Waters, Neeans, NE of West Burrafirth (1989, WS); swamp by road W of the N end of Grass Water; and at the Loch of Houster, Aith. One site is known in North Mainland: Loomi Shun, S of Kirkhouse Loch, Vidlin (1996, R. C. Palmer). There is also one station in Northmavine, SE of Sand Water, SE of Tingon (2002, B. O'Hanrahan). In Fetlar it occurs by the Burn of Funzie; also at the Loch of Stivla (1991, M. Tickner). There are two sites in Unst: W of Haroldswick, on both sides of the turn to Ungirsta; a small patch E of the Burn of Mailand, at HP602076 (2003, C. Geddes and A. Payne).

C. aquatilis. *Water Sedge*. Very rare. Discovered in 1967, this sedge is known only from the junction of the Burn of Northdale with Papil Water, Fetlar. It was in good condition in 2003 (R. Norde).

C. nigra. *Common Sedge*. A very common, widespread and variable sedge of a range of damp or watery habitats from marshes and bogs, margins of lochs, streams, and ditches, to heaths and moors where it often grows among heather.

C. bigelowii. *Stiff Sedge*. A frequent arctic-alpine sedge of the higher hills where it grows in damp or dry, often mossy, moorland (including peat-haggs), on fellfield and associated *Rhacomitrium*-heath on Ronas Hill, Northmavine, as well as on stony summit plateaux elsewhere. Stiff sedge also occurs very locally at much lower levels, as on the dry stony heaths on the North Roe plateau, N of Ronas Hill, on stony-mossy heaths on the higher parts of the Hill of Colvadale and Sobul, both in Unst; on a few moory holms in lochs; and to as low as fifteen m at Tressa Ness, on the N side of Fetlar.

C. pulicaris. *Flea Sedge*. Common and widespread in flushes and on wet rocks, notably in limestone and serpentinite areas; also in damp heathy pastures, on stony loch shores, etc.

Nardus stricta. *Mat-grass*. An abundant and widespread grass on damp heaths and moors, often with *Juncus squarrosus*, and in damp peaty grassland.

*****Festuca pratensis**. *Meadow Fescue*. An occasional and well-established alien of the crofting areas, especially by grassy roadsides, margins of hayfields, etc. Long ago this may have been a component of grass-seed mixtures; nowadays it is more likely to be a contaminant of such mixtures.

*****F. arundinacea**. *Tall Fescue*. A very rare and weakly established alien. First recorded, in 1956, by the shore road, Haroldswick, Unst, and still there in 2002 (L. A. Inkster and WS). In some quantity on both sides of the A970 near the Loch of Clickimin, Lerwick, 1964, but by 1969 it had been largely destroyed by roadworks. In 1982 two clumps were seen by the nearby Sea Road, close to its junction with the

A970; one clump persisted in 1991, but by 1999 it could not be found (R. C. Palmer). A large clump was found in 1991 by the B9074 a little S of the standing stone N of Asta, near Scalloway (RCP); in 2008 it was still present (WS). In 1996 RCP noted it twice in South Mainland: several large patches by the track between Sumburgh Farm and Sumburgh Hotel, still there (2008, WS), and a large clump by the E side of the A970 near Exnaboe.

F. rubra. *Red Fescue*. A widespread and variable grass which occurs abundantly in the county. Seven subspecies are recognised in Britain and all—five native and two alien—are recorded from Shetland. Subsp. **rubra** is by far the commonest and grows in meadows and pastures, on coastal ciffs and banks, and on sand-dunes, rocks, and by roadsides, etc. Subsp. **juncea** has been found, always as the pruinose form, in a few rocky coastal sites as well as in a handful of more inland places, as on wet stony slopes above the Loch of Haggrister, Northmavine. Subsp. **litoralis** is on record from the saltmarsh at Hoove, Whiteness, Central Mainland, and may occur in this and other coastal habitats elswhere. Subsp. **arctica** is widespread on the Unst serpentinite from Crussa Field to Muckle Heog, both on fellfield and in rock-crevices; also on serpentinite gravel by the roadside at the Small Waters, NNW of Uyeasound (1989, R. C. Palmer). The 2006 BSBI field meeting found it on serpentinite on the Keen of Hamar and on the Hill of Clibberswick, both in Unst; on granite on Ronas Hill, Northmavine, but very sparingly; and in turf bordering the beach, Effirth, West Mainland. All the field meeting material was identified by A. Copping, who was also a member of the meeting. Subsp. **scotica** is a very rare British plant with only ten records, mainly from Scotland (Preston, Pearman & Dines 2002), including the only one from Shetland: low rocky sea-cliffs, S side of Firths Voe, near Mossbank, North Mainland (1976). *Subsp. **commutata** (Chewing's fescue), a casual, was found by roadsides near the Loch of Urafirth, Northmavine, in 1979 as a grass-seed component or contaminant. *Subsp. **megastachys**, another casual of similar status, was found in 1966 by a cornfield near Scalloway and on the local rubbish-dump. In 1996 RCP noted it in two places in South Mainland (sown slope at Williamsetter and on the sown verge of the A970 near Crossie Geo, Sandwick), as well as on a disturbed roadside above Gerda Water, E of Vidlin, North Mainland. Much more research is needed to establish the habitat requirements of our four apparently rare or local native subspecies, subsp. *juncea, litoralis, arctica,* and *scotica.*

F. vivipara. *Viviparous Sheep's-fescue*. A widespread and common grass of the poorer soils: heaths, moors, turfy banks, among rocks, and on fellfield. Although often found in dry places it will also occur in wetter areas such as the drier parts of bogs, sides of peaty ditches, etc.

× **Festulolium loliaceum** (*Festuca pratensis* × *Lolium perenne*). *Hybrid Fescue.* An apparently very rare hybrid with only two old records: grassy pastures between Norwick and the sea, Unst (1956), and in a field-border at the Wick of Gossabrough, Yell (1961).

*****Lolium perenne**. *Perennial Rye-grass*. Despite some very early references to the indigenousness of this species, WS regards it as a well-established alien which has been imported into Shetland over a very long period as a vital component (along with *Trifolium repens*) of seed mixtures for surface seeding, hay, etc. Spence (1914) regarded it as having been intoduced in Orkney. It is common in Shetland, especially on trampled ground around crofts and farms. Its occasional appearance in rocky limestone grassland is no guarantee of its nativeness in these areas.

*L. **multiflorum**. *Italian Rye-grass*. Casual. A mere relic or waif of cultivation.

*L. **temulentum**. *Darnel*. A very rare casual on the Scalloway rubbish-dump (1961).

***Vulpia bromoides**. *Squirreltail Fescue*. Another very rare casual, this time with three old records: in and by a hayfield, Bouster, Yell (1961); turfy slopes (probably once cultivated) above the croft of Sandwick, near Swining, North Mainland (1963); and one tuft by wall, junction of New Road and Castle Street, Scalloway (1963).

Cynosurus cristatus. *Crested Dog's-tail*. Common and widespread in dry or slightly damp pastures, on grassy banks and in rocky grasland, especially on fertile soils.

*C. **echinatus**. *Rough Dog's-tail*. A very rare casual which was seen twice in the same area within the space of a few years in its only recorded station: Bressay, about eight specimens on sandy ground near the S side of the Noss road in October 1840; in July 1843 three small plants were seen about 100 yards from the original site.

Puccinellia maritima. *Common Saltmarsh-grass*. An occasional to frequent grass. Characteristic of saltmarshes, brackish turf, and quiet muddy seashores (sometimes inundated at high tide); rarely on sandy or shingly beaches, small holms or in coastal rock-crevices.

P. **distans** subsp. **borealis**. *Reflexed Saltmarsh-grass*. A common and widespread grass of coastal habitats: sea-cliffs, stacks and skerries (where it is often fouled by sea-birds), seashores (both shingly or muddy), saltmarshy turf (rarely), and on old coastal stonework. In 1991 L. A. Inkster and WS noted that this was the most northerly flowering plant in Britain. It grew on the N side of the Out Stack, N of Unst and was, admittedly, only a few inches N of *Cochlearia officinalis* subsp. *officinalis*. Perhaps their roles are reversed in some years.

Briza media. *Quaking-grass*. Very rare as an almost certain native in rich damp pastures by the SW side of the Loch of Tingwall, Central Mainland (where it has occurred sparingly since 1956), and in a similar situation above the SW corner of the Bay of Ollaberry, Northmavine, where it was found in abundance in one meadow in 1993 (P. V. Harvey) and very sparingly in 2004 in two neighbouring meadows (PVH). There are nine other records which WS regards as references to casual occurrences, while being mindful that one or two could refer to native stations. These are as follows: Shetland, unlocalised, *c*.1908; West Mainland, unlocalised (1991, A. J. Silverside); a small patch in a field close to Halligarth, Baltasound, Unst, *c*.1906 (I. M. F. Sandison, per I. D. J. Sandison), well over 100 plants, but surrounded by a thick growth of *Holcus lanatus* (2000, M. G. Pennington); roadside verges and ditch banks at the junction of the B9074 and the (old) A970 near the Scord Quarry, Scalloway (1958), still present but showing little if any tendency to spread; roadside near the N end of the Loch of Tingwall, Central Mainland, *c*.1960; by the A970 near the Loch of Huesbreck, and by the roadside near Outvoe, Boddam, both South Mainland (both 1965); pasture by stream, Brouster, near Bridge of Walls, West Mainland, about fifteen plants (1965); and two plants in limestone pasture SE of the old lime kilns, Laxfirth, Tingwall, Central Mainland (1988, D. H. Dalby).

***Poa annua**. *Annual Meadow-grass*. An abundant and widespread grass. Its chief habitats are closely associated with human activity (arable and garden ground, gravelly paths and roadsides, walls and foreshores, and open disturbed ground generally. This makes it difficult not to regard it as a well-established alien, despite its currently accepted national status as a native. Annual meadow-grass is also found in more natural surroundings, such as on sea-cliffs and holms in lochs; here, it may have been introduced by birds from local sources.

P. trivialis. *Rough Meadow-grass.* A common and widespread grass which is found mainly in low-lying cultivated districts. Damp grassy places, burns and ditches (often with *Iris pseudacorus*), arable land, open waste ground, shingly seashores, etc.

P. humilis. *Spreading Meadow-grass.* The third in a trio of common and widespread meadow-grasses. Unlike the preceding, this species prefers much drier places: dry (often rocky) grassland; sandy coastal pastures and among sand-dunes; shingly beaches; parched mossy-turfy wall-tops of long-abandoned crofthouses; also (but less commonly) in damp meadows and saltmarshy turf.

***P. pratensis.** *Smooth Meadow-grass.* Casual. A component of grass-seed mixtures and thus occasionally recorded from waste ground, foreshores, grassy enclosures, etc. Probably widely overlooked as it can often resemble large forms of *P. humilis*.

***P. flabellata.** *Tussac-grass.* Casual. This is a very long-persistent grass which was deliberately introduced long ago around crofts, usually inside stone-walled yards, in South Mainland—Fladdabister being a well-known station. There is also a good example at Ham, Bressay (see below). In 1847 one early grower in Shetland, perhaps the only one, and possibly 'Mr Bruce, of Sumburgh' (Gorrie 1855), had a plant five feet high. The writer is indebted to L. F. Anderson for pointing out that seeds from the Falkland Islands (one of its native strongholds in the South Atlantic) were sown at Ham, Bressay, in 1899; the success of that planting can be seen to this day. See Anderson (2003). Elsewhere in Shetland it is very rare, but at one time it used to grow by the forecourt of the Lerwick Hotel. Technically, it scarcely merits mention in a local flora as few if any of its stations are totally enclosed. However, it is a large plant which has been around for a long time; this, along with its flowering from late February to April, are points of interest, if nothing more.

***Dactylis glomerata.** *Cock's-foot.* An important component of grass-seed mixtures, this is a common, widespread and well-established alien of rough grassy places: roadsides, field-borders and neglected corners of arable fields, waste rubbly ground, and on or by old walls.

Catabrosa aquatica. *Whorl-grass.* Very rare but sometimes locally abundant. Apart from three old records by Edmondston (1845), this aquatic grass is known from two areas only, the S part of South Mainland and the N end of Yell where, in both places, it enjoys rich muddy or sandy-watery places. In the first area its distribution is summarised by Scott, *et al.* (2002), 'In South Mainland it is found in sluggish ditches and streams, in marshes and muddy hollows, and by sandy loch margins from behind the Bay of Quendale, through the Loch of Hillwell and the Links of Quendale northwards to the Bakkasetter wetlands and the S and W sides of the Loch of Spiggie.' To the foregoing could be added the Loch of Brow area. It was also seen near Toab (1924), and in a small flooded area by runway 15/33, Sumburgh Airport, near the S corner of Grutness Voe (1966). In the N of Yell it grows in the sandy stream from Kirk Loch, and in muddy places in hayfields E of Breakon. The three old records referred to are: near the manse at the N end of the Loch of Tingwall, Central Mainland; pool on the E side of Whalsay; and near Lund, Unst. All of these were made in the late 1830s or early 1840s by T. Edmondston. None has been confirmed, but the fairly extensive sands at Lund would appear to make it a likely area.

Glyceria fluitans. *Floating Sweet-grass*. Frequent and widely spread throughout the county in watery places, both in peaty or (more commonly) eutrophic situations. Shallow water in lochs and pools where it commonly floats on the surface; sluggish ditches and burns in agricultural areas, often luxuriant through habitat enrichment; muddy hollows and swampy places, etc.

[**G. declinata**. *Small Sweet-grass*. Believed to be extinct. There are two records: in a muddy depression in a field by the Burn of Weisdale, below Kergord, Weisdale, Central Mainland (1966); on wet silt at the N end of the Loch of Vatsetter, Yell (1994). Despite searches at both places the grass has not been seen again.]

*****G. notata**. *Plicate Sweet-grass*. A very rare casual. One record: one clump by the Unst Leisure Centre (1989). Probably a grass-seed contaminant.

Helictotrichon pubescens. *Downy Oat-grass*. An occasional grass of dry places, especially on limestone or sandstone soils: pastures (often rocky or hummocky), dry grassy banks, low sea-banks, streamsides, and grassy rock-ledges. Widely but thinly distributed, and unrecorded from Yell, Fair Isle, Foula, Bressay, and Whalsay.

*****Arrhenatherum elatius**. *False Oat-grass*. This seems to be a well-established alien. Frequent and often common, especially on the better soils and in agricultural and inhabited areas. Rough grassy places, cultivated and waste ground, roadsides and associated ditch banks, low rubbishy sea-banks and the higher parts of shingly beaches. Var. *elatius*, and var. *bulbosum* (onion couch), are both found in Shetland. Although both varieties share a range of habitats, the latter is mainly a feature of cultivated ground and was, at least formerly, a troublesome weed.

*****Avena strigosa**. *Bristle Oat*. A very rare casual as a relic or outcast of cultivation. These occurrences are few and far between as *A. strigosa*—in the distant past the only oat grown in Shetland—is now seldom cultivated, its place having been taken by *A.sativa* (oat). Fortunately, it has not quite vanished; it is still (2008) grown in Fair Isle for use in the making of Fair Isle straw-backed chairs. In a few other places (notably Burland, Trondra) it is grown for small-scale use as well as for its historical interest, along with the need to ensure that the Shetland stock of this almost forgotten cereal is not lost. Although bristle oat is also a scarce casual elsewhere in Britain, its Shetland occurrences are likely to relate in large measure to its former cultivation in the islands.

*****A. fatua**. *Wild-oat*. A rare casual with no recent records. Its habitats, where noted by the recorders, were cultivated fields of barley or oats, weedy roadsides and rubbly places. Apart from an unlocalised notice (1837 or earlier) there are records from the following stations. Sandwick, West Burra (1966). West Mainland: Culswick (1960, 1981); Bevla, Wester Skeld (1967); West Burrafirth (1980). Northmavine: no locality or localities given (before 1845); Hamnavoe (1880). Yell: Gossabrough (1804). Unst: Burrafirth (1837); Baltasound pier (1977).

*****A. sativa**. *Oat*. Casual. An outcast of cultivation which is occasionally seen around crofts and farms, by the edges of cultivated fields, and on neighbouring rubbishy sea-banks, foreshores, etc. Long ago it almost completely replaced *A. strigosa* as the chief oat crop in Shetland. However, it is itself much less cultivated now than in the past, owing to changing agricultural practices.

*Trisetum flavescens. *Yellow Oat-grass*. A rare casual of roadsides, even rarer in gardens, usually in small quantity and with no recent sightings. Haroldswick, Unst (1956). Hillswick, Northmavine (1958, 1979). South of Asta, near Scalloway (1958). Garden, New Road, Scalloway, in grass (1962); and at Uppersound, Lerwick (1959).

Deschampsia cespitosa. *Tufted Hair-grass*. Common and widespread in moist grassy places: meadows; pastures; turfy, heathy ground; sides of burns; grassy sea-banks; holms in lochs; and by roadsides, etc.

D. setacea. *Bog Hair-grass*. A scarce grass, both nationally and locally. In Shetland it is known from one area only—North Roe, Northmavine—where it is very thinly and intermittently scattered over a fairly wide area bounded by Ure Water in the NE, the Queina Waters in the SE, near Swabie Water and Sandy Water in the SW, and, in the NW, by the Burn of Moosawater. It is a grass of watery, peaty or stony places, sometimes in small swampy pools where it is submerged apart from the inflorescence. It is most likely to be encountered in the following one-kilometre squares: HU3384, 3385, 3484, 3485, and 3585. Even here it is likely to be represented by a handful of widely separated plants (rarely upwards of a hundred) in any one place. The writer is indebted to L. A. Inkster for undertaking, in 2001, a survey of this rather restricted species.

D. flexuosa. *Wavy Hair-grass*. Common and widespread, often locally abundant. Heaths and moors, from dry to wet and mossy; rocky-heathery places among the hills or by burns; on fellfield on Ronas Hill, Northmavine, and other stony areas on or near the summits of the higher hills; and on peaty holms in lochs.

Holcus lanatus. *Yorkshire-fog*. One of our most common and widespread grasses, often locally abundant. Meadows and pastures (wet, dry, or sandy), damp heaths, marshy areas, rough grassy places, holms in lochs, sea-banks, etc.

*H. mollis. *Creeping Soft-grass*. A well-established alien which is chiefly found by roadsides and associated ditch banks. Other habitats include grassy streambanks, waste grassy ground, and gardens (where it occurs as a weed, and occasionally under planted trees and shrubs). It is rare in South Mainland and Northmavine, and local to occasional in Central, West and North Mainland, Whalsay, Fetlar, Yell, and Unst. In Bressay in 1991 it was common in the more populous parts of the island.

Aira caryophyllea. *Silver Hair-grass*. Rare and perhaps decreasing. Probably native in some stations, possibly casual in others. Dry grassy, sometimes rocky or shaly, banks both inland and by the coast; sand-dunes (as, at least formerly, on the Links of Quendale, South Mainland); roadside banks; and (rarely) turfy slopes which may once have been cultivated. Its fourteen records are distributed as follows: South Mainland (3), Central Mainland (5), West Mainland (1), North Mainland (3), and Unst (2). It was last noted, in 1989, at Burra Firth, Unst (R. C. Palmer).

A. praecox. *Early Hair-grass*. A very common and widespread grass of dry places, especially in turfy heathy grassland, on bare or stony peat and associated peat-haggs, among rocks, on parched mossy-turfy wall-tops of derelict crofthouses (often with *Poa humilis*), and on open waste ground, etc.

Anthoxanthum odoratum. *Sweet Vernal-grass*. An abundant and widespread grass in meadows and pastures, on heaths and moors; particularly associated with dry heathy grassland. Edmondston (1809) correctly observed that it 'communicates that fine flavour which hay exhales shortly after it has been mown'.

Phalaris arundinacea. *Reed Canary-grass.* A common, widespread and frequently abundant grass of mainly watery or damp places, especially on the better soils in the lowlands: in meadows, by burnsides and ditches, and on waste ground, etc. Two somewhat unusual habitats may be cited: on low sea-banks E of the old house on Bigga, Yell Sound (2003, WS), and on the holm in the Loch of Clickimin, Lerwick, the holm in the Loch of Huxter, Whalsay, and the holm in the S end of the Loch of Cliff, Unst (all 2003, WS). A cream and green striped garden form is occasionally seen as an outcast or where it has been deliberately introduced. It can be very persistent, as by a burn at Houbie, Fetlar, where it was first recorded in 1928 but may not now be surviving.

*P. canariensis. *Canary-grass.* A rare casual of gardens, rubbish-dumps, foreshores, etc., probably mainly originating from bird-seed mixtures, and certainly so in some cases.

Agrostis capillaris. *Common Bent.* Common and widespread on dry, often stony, heathy pastures, damp grassy moors, turfy places among rocky outcrops, etc.

A. × murbeckii (*A. capillaris* × *A. stolonifera*). Once recorded, from high ground above the Daal Burn, Foula (1956).

*A. gigantea. *Black Bent.* A very rare casual which has been recorded from only two areas, both in the S of South Mainland, and not recently. Near Boddam, habitat not noted (1960); weed in sandy arable fields, Exnaboe (1966, and still there, very sparingly among barley, in at least one place in 1982).

*A. castellana. *Highland Bent.* An even rarer casual. One record: on bare soil on the sea-bank below the North Isles Motel, Sellafirth, Yell (1997, R. C. Palmer). Included in grass-seed mixtures and probably much overlooked in Shetland. It is not always easily separable from *A. capillaris*.

A. stolonifera. *Creeping Bent.* Frequent to common in a wide variety of habitats. Among rough herbage; in marshy pastures and by ditches and burns (sometimes with *Iris pseudacorus*); also by the coast in saltmarshes and brackish turf, on coastal sands, cliff-tops, holms and stacks; on the serpentinite heaths of Unst and Fetlar; and in or by cultivated land, on waste ground, and by roadsides, etc.

A. vinealis. *Brown Bent.* Another common and widespread bent. This is a species of wet or dry places: grassy moorland, heaths, bogs and swamps. The closely related *A. canina* (velvet bent), a species of damp or wet habitats throughout most of the country, appears to be absent from Shetland despite the abundance of seemingly suitable habitats.

Ammophila arenaria. *Marram.* Sand-dunes and associated dune-pastures and the higher parts of sandy beaches adjoining dunes; also on low grassy sea-banks by or near sandy or (rarely shingly) beaches in the absence of a dune system. Frequent in the S part of South Mainland, especially the area behind the Bay of Quendale where marram has its most extensive site in Shetland, and where it is abundant and extends well inland. Elsewhere in the county it is rare or local. Twenty-five stations are on record, distributed as follows: South Mainland (9); Isle of Noss (1); Whalsay (1); below Sandwick (2005, L. A. Inkster and WS); West Mainland (3); Papa Stour (2); Northmavine (1), at Brae Wick (1999, WS); Yell (2); Fetlar (1); Unst (4), and Uyea, off Unst (1). It is sometimes seen near houses and by roadsides, etc., as an accidental introduction in sand from the coast, occasionally persisting for quite some time.

***Apera spica-venti**. *Loose Silky-bent*. A very rare casual. Three plants, as an almost certain grass-seed contaminant, by the Unst Leisure Centre (1989).

***Alopecurus pratensis**. *Meadow Foxtail*. A well-established and frequent alien of grassy places in the inhabited or cultivated areas except West Mainland where it seems to be scarce. Waysides, field-borders, near churchyards and houses, especially on the more productive soils. This was probably once included in grass-seed mixtures, either intentionally or as a contaminant.

A. × brachystylus (*A. pratensis × A. geniculatus*). Very rare. Once collected: damp grassland in the valley above the Gulberwick church, Central Mainland (1979).

A. geniculatus. *Marsh Foxtail*. A frequent to common grass throughout the county in ditches and burns, by loch margins, in wet meadows and pastures (often flooded in winter), and in damp arable ground.

***A. myosuroides**. *Black-grass*. A very rare casual. Once found (in 1960) on the Lerwick rubbish-dump (on the site of the North Loch).

***Phleum pratense**. *Timothy*. Casual. This is an important component of grass-seed mixtures for hay, and for the reseeding of newly prepared ground, such as roadside embankments, etc. It is thus a fairly frequent relic of cultivation by roadsides and field-borders, about crofts and farms, and on waste rubbly ground.

***Bromus commutatus**. *Meadow Brome*. A very rare casual with only two records: one plant on a rock by the Burn of Dale, above Netherdale, West Mainland (1980); common in a hayfield between Williamsetter and Ireland, South Mainland (1981). Probably a grass-seed contaminant in both cases.

***B. hordeaceus** subsp. **hordeaceus**. *Soft-brome*. A well-established alien which is found mainly in low-lying cultivated and inhabited areas: in weedy neglected fields and by their borders, sides of roads and tracks, etc. This grass is fairly distributed in the county, and is recorded from thirty-three ten-kilometre squares. However, only seventeen of these are from 1987 or later, and this supports the writer's opinion that soft-brome is probably decreasing in Shetland.

B. × pseudothominei (*B. hordeaceus × B. lepidus*). *Lesser Soft-brome*. A very rare casual which probably arose spontaneously. There are only four records, all old, and from hayfields unless otherwise indicated. Seli Voe, near Gruting, West Mainland (1959); Lunning, NE of Vidlin, North Mainland (1960); near the mouth of the burn from Uradale, SE of Scalloway (1962); and on the shingle beach, Papil, West Burra (1978).

***B. lepidus**. *Slender Soft-brome*. Nowadays a casual if still extant. An intentional constituent or contaminant of grass-seed mixtures, first recorded in 1955 from near the Loch of Asta, Tingwall, Central Mainland. This was once a well-established but short-lived alien of hayfields, waste ground, roadsides, etc., which, from the mid 1950s to the mid 1960s, was occasional to frequent (sometimes abundant) in the S half of Mainland (especially South Mainland), and in parts of Unst, Yell, and Fetlar. It decreased rapidly in the late 1960s; indeed, its last sighting was in 1968 at Walls, West Mainland, by R. C. Palmer. The date of its arrival in Shetland is not known; it was not noted in the early 1920s by G. C. Druce.

***B. pseudosecalinus**. *Smith's Brome*. A formerly very rare casual of hayfields: Gruting, West Mainland (1959); Ham, Bressay (1959). It is likely to have been a grass-seed contaminant.

94

*__Anisantha sterilis__. *Barren Brome*. This very rare casual is known from only three stations, all in Central Mainland. Garth, South Nesting, rockery weed (1982, still there in 1986); between Westshore and Port Arthur, Scalloway, a number of plants on a reseeded roadside verge (1992); in garden of 9 Knab Road, Lerwick, of bird-seed origin (1994).

*__Elytrigia repens__ subsp. __repens__. *Common Couch*. Status unclear, but it seems more likely to be a well-established alien than a native. A widely distributed grass of cultivated land, margins of fields, roadsides, rubbish-dumps and other open rubbly ground, low sea-banks and foreshores, and a troublesome weed in gardens.

__E. repens__ × __E. juncea__ (*E.* × *laxa*). A local plant of sandy beaches and adjacent sandy banks, rarely on shingly seashores; sometimes with one or both parents missing. Of the nineteen recorded stations, four are in Unst, five in Yell, and one in Fetlar. Included in these is one new site: on gravel by the harbour, Uyeasound, Unst (1991, L. A. Inkster and WS). There is one site in Central Mainland, two in West Mainland, one site in North Mainland, and three in Northmavine. It is also on record from West Burra and Whalsay. There are no records at all from South Mainland, despite the number of suitable habitats and the presence of both parents in the S part of this division.

__E. juncea__ subsp. __boreoatlantica__. *Sand Couch*. A local to occasional (sometimes abundant) grass of sandy or sandy-shingly seashores, and on the seaward edge of sand-dunes where it often grows with *Ammophila arenaria* (marram) and *Leymus arenarius* (lyme-grass), the three being essential for the formation and maintenance of a dune system. Rarely seen on low rocky or sandy sea-banks behind a sandy beach. It is known from twenty-nine stations. Ten of these are in South Mainland, mainly from the S part; two in Central Mainland; two in West Mainland, and one each in North Mainland, Northmavine, Fetlar, West Burra, and the Isle of Noss; four in Yell; five in Unst; and one in Huney, off Unst (1989, R. C. Palmer).

__Leymus arenarius__. *Lyme-grass*. A frequent and widely distributed grass of coastal habitats. Sandy beaches (with or without dunes), also on shingly or bouldery beaches, steep-sided stacks, holms and *geos*, and low rocky sea-banks. In these places it is often in small patches. However, on sand-dunes and their associated dune-pastures and sandy beaches, this sand-binding grass is often plentiful, growing with and sometimes replacing *Ammophila arenaria* (marram).

*__Hordeum vulgare__. *Six-rowed Barley*. A very rare casual. Now almost extinct as a crop, but still being grown in a few places (as at Burland, Trondra) where it occurs as a relic or outcast of cultivation. In 2008 it was abundant in two fields of *H. distichon* (two-rowed barley) between Scalloway and Utnabrake (WS), a casual of local origin on this occasion. Nevertheless, its possible presence elsewhere as an accidentally imported casual cannot be ruled out as it is an occasional casual in other parts of Britain. In our distant past this, along with *Avena strigosa* (bristle oat), were the only cereals grown in Shetland. Both are now faint memories of an earlier period in our crofting and farming history.

*__H. distichon__. *Two-rowed Barley*. Casual. A very rare casual as a relic or outcast of cultivation, mainly in cornfields, but liable to appear on beaches, as in Colsay, off South Mainland, where a few plants along with a similar number of *Avena sativa* (oat) grew on a beach on the W side of the island (2002, WS).

*H. jubatum. *Foxtail Barley*. A rare casual of reseeded ground—a grass-seed contaminant and usually in small quantity—which appeared in a number of places in the 1980s but has not been recorded since. Central Mainland: South Lochside area, Lerwick (1982, 1985): Unst (all in the Baltasound area), as follows; small patch on croft above the Burn of Mailand, W of Trolla Water [near the school] (1984); among houses at the head of Balta Sound (1984); by or near the Unst Leisure Centre (1989, R. C. Palmer); in a field, Hillsgarth (1989, R. H. W. Anderson). North Mainland: a single plant on the sown verge of the A970 ENE of Hultness, near Wethersta, and close to where it crosses the Mill Burn (1991, RCP). In Trondra it has been found by a new house, 'Aerisdale' (1989, E. K. Sinclair).

*Triticum aestivum. *Bread Wheat*. A rare casual of rubbish-dumps, arable land, gardens, etc. It is sometimes seen as a bird-seed alien as at 9 Knab Road, Lerwick (1994, I. Clark). There are records from South and Central Mainland, and a very old record from Bressay by P. Neill in 1804 from the W shore of the island.

*T. turgidum. *Rivet Wheat*. Casual, very rare. One plant, Scalloway (1976).

Danthonia decumbens. *Heath-grass*. A frequent to common species of dry or wet (often mossy or turfy) heathy grassland, grassy banks, rocky streamsides, etc.

Molinia caerulea. *Purple Moor-grass*. A common and widely distributed grass of wet heaths and moors (especially on the serpentinite and greenstone of Unst and Fetlar); basic flushes; marshes and other base-rich watery situations; stony heathery mires; and on holms in lochs and other sheltered ungrazed places where it is much taller than on the neighbouring moorland.

Phragmites australis. *Common Reed*. Rare and perhaps gradually declining. There are eleven recorded sites scattered over the county, but in one of these it is almost certainly extinct and in one other it has not been observed recently. Marshy or wet burnsides; ditches; damp grassland; swampy places; and rarely in small relatively rich lochs. In one place only (Clavel) it forms large pure stands, but elsewhere it occurs in smaller colonies or as scattered groups. In South Mainland there are four recorded sites: stream below Clavel, abundant over a large area; near Sand Lodge, Sandwick, first and last seen in 1868 and probably extinct; lower end of the Burn of Mail, Cunningsburgh; by the Burn of Laxdale, between the A970 and Stanefield (2007, WS). Central Mainland: by the W side of the Vadill of Garth, South Nesting; SW of the Brettabister church, very small scattered plants (2008, WS) where in 1979 there was a large bed of it; near and in roadside ditches NE of the Brettabister church, two occurrences, the larger 100 m long (2008, WS), the two last-named sites in North Nesting. It is in good condition in the small loch near Hogan, below the road to Snarraness, its only West Mainland locality. North Mainland, one station: at the SE corner of the head of Dales Voe, Delting. Northmavine, one station: frequent near a large marshy-swampy area S of 'Findlins House', Ness of Hillswick, all plants very short and barren (1987, WS), not found in 2008 (WS). In Unst there is a large patch by the Burn of Mailand, near Baliasta.

*Panicum capillare. *Witch-grass*. A bird-seed casual. Very sparingly near new roadworks, 'South Taing', Cott, Weisdale, Central Mainland (2009, R. Leask).

*Setaria sp. A bird-seed casual. Very sparingly on disturbed ground near new roadworks, 'South Taing', Cott, Weisdale, Central Mainland (2009, R. Leask). This is almost certainly **S. viridis** (green bristle-grass), according to E. J. Clement.

Sparganium erectum. *Branched Bur-reed*. Very rare. As a native it is known from one site only, and in very small quantity, but is abundant as a well-established alien in another location. In 1978 two large patches were discovered among *Iris pseudacorus* by the swampy margin of the brackish pool near Scarfa Skerry, Ness of Sound, Lerwick. During the following years the plant deteriorated and by the mid 1990s it existed as a mere shadow of its former self. In 1996 L. A. Inkster and WS—in the apparent absence of any meaningful attempt to prevent it from being lost to Shetland—transplanted a few rhizomes to the S margin of the nutrient-rich Loch of Hillwell, South Mainland. It proved to be a very successful operation, and one which secured the continuation within our county of a very rare plant by using original native stock. Happily, its decline at the Lerwick site seems to have been halted. In 2008 WS found two tiny clumps below the road and four sizeable patches (some with fruiting plants) by the E side of the pool. However, the plant is still very vulnerable at Lerwick and deserves sustained attention from conservationists. Its abundance as an *alien* at the Loch of Hillwell should not lead to its disregard at its only *native* station. The Shetland plant appears to be closest to subsp. **neglectum**.

S. **angustifolium**. *Floating Bur-reed*. This is an occasional to frequent and widely distributed plant of shallow or deep water in pools, lochs, and slow-flowing burns, rarely in swamps; chiefly in hilly moorland areas, less often in mildly nutrient-rich waters. It sometimes flowers abundantly, but at other times produces leaves only.

S. **natans**. *Least Bur-reed*. This very rare plant occurs somewhat sparingly in a deep meandering burn in Unst—the Burn of Caldback, in a section NNW of the Black Loch. In former times it was recorded more widely in this burn, and it is suspected that drainage activities may have had a detrimental effect. The future safeguarding of this rare and apparently declining rarity from potentially damaging operations should be of the utmost concern to those who are empowered to address conservation matters of this kind.

*****Typha latifolia**. *Bulrush*. Casual, almost certainly planted. About nine plants in a wet pasture just below the road below Bardister, Northmavine (2004, A. Williamson). By 2008 it had increased to twenty-five plants, of which several had flowered (WS).

Narthecium ossifragum. *Bog Asphodel*. Common and widespread on wet heaths and moors, and in swampy-peaty places. A form with pale greenish-yellow flowers has been recorded twice: a colony S of the new school, Brae, North Mainland (1982, J. S. Blackadder); the site was destroyed in 1993 during the construction of a sports field. About sixteen plants *c.*100 m W of the house, 'Hole in the Wall', Bridge of Walls, West Mainland (2006, S. Beer). Growing near or with the type in both cases.

*****Hosta** sp. *Plantain Lily*. Casual. Merely a rare garden outcast. Many cultivars and hybrids are available through the horticultural trade.

*****Colchicum autumnale**. *Meadow Saffron*. Casual. Another garden plant which has occurred very rarely as an outcast or as a deliberate introduction.

*****Lilium pyrenaicum**. *Pyrenean Lily*. Casual. A hardy and very popular garden plant in the county, and one which is sometimes seen as an outcast (by burns, on waste ground, etc.), or as a deliberate introduction.

*****Ornithogalum angustifolium**. *Star-of-Bethlehem*. A very rare casual of garden origin. Three records: in an old quarry by the road near Bakkasetter, South Mainland (1992); here and there about houses, Clett, Whalsay (1996); and several clumps (probably planted) on the grassy verge by the N side of Ingaville Road, Scalloway (1995), still present in 2010, but grass-cutting operations severely limit its chances of flowering.

Scilla verna. *Spring Squill.* A common, widespread, and often abundant plant of dry grassy pastures near the sea, less often inland; maritime heaths; small grassy offshore holms (often with *Armeria maritima* subsp. *maritima* (thrift) and soaked in the spray of summer gales); grassy ledges and niches among rocks; and on the serpentinite heaths and associated debris of Unst and Fetlar. Spring squill is a welcome reminder of approaching summer.

***Hyacinthoides non-scripta* × H. hispanica**. *Bluebell.* A well-established alien and one of the most popular garden plants in Shetland. It is frequently seen as an outcast near houses, churchyards and cemeteries, and on waste ground and low rubbishy sea-banks, etc., but is probably a deliberate introduction when found at some distance from habitation. The hybrid has recently been published as **H. × massartiana**.

Allium ursinum. *Ramsons.* A long-persistent casual in one place: among and in the the vicinity of the remains of the ruined house at Sound, Weisdale, Central Mainland.

Narcissus. *Daffodil.* Along with the bluebell this is another well-established alien of garden origin which is often seen as an outcast in grassy places about houses and crofts. It is also planted by new roadside verges and embankments. Many gardens contain an assortment of cultivars. An old cultivar, not uncommon around the ruins of long-abandoned crofthouses, has numerous perianth-segments which are greenish at first but become fully yellow at maturity.

Alstroemeria aurea. *Peruvian Lily.* Casual. This is often grown in gardens and is thus sometimes found as a persistent outcast, escape, or deliberate introduction.

Iris pseudacorus. *Yellow Iris.* A common and widely-distributed plant of marshes and bogs on all kinds of soil, especially by burns in low-lying valleys where it often forms large colonies, and where it may be waterlogged for long periods in wet winters. In some seasons it flowers abundantly, in others very few blooms are produced.

I. latifolia. *English Iris.* A casual of garden origin, sometimes seen as an outcast in grassy places near houses or where it has been deliberately introduced.

***Gladiolus* sp**. *Gladiolus.* Casual. Planted [in 1998] in a few scattered locaities in the inhabited area, Fair Isle (N. Riddiford). A small early-flowering form which by 2008 had disappeared from all but one site. The writer is indebted to N. Riddiford and B. Skinner for information concerning this alien.

Crocosmia paniculata. *Aunt-Eliza.* Casual of garden origin. Bressay, a garden outcast or deliberate introduction by the W side of the shore road in two places: in front of the Voeside houses, and *c.*150 m SSW of the turning to Glebe Park (both 1998, WS). See note under the following species.

C. masoniorum. *Giant Montbretia.* Casual of garden origin. Sparingly among the rock-armour by the B9074, adjacent to the marina at the E side of the East Voe of Scalloway (2008, WS). This and the preceding species have both gained popularity with Shetland gardeners in recent years and can be expected to appear more widely outside cultivation, and to become established aliens in the years to come.

C. × crocosmiiflora (*C. pottsii* × *C. aurea*). *Montbretia.* A well-established alien of garden origin. As an outcast or deliberate introduction it occurs frequently by burns (as by the Burn of Mail and the Burn of Laxdale, both in Cunningsburgh, South

Mainland), on steep grassy sea-banks (as by the NE corner of Brei Wick, Lerwick), and by ditch banks and roadsides, etc., in and near inhabited areas. It also fills a *plantie-crub* at Bankwell, at the S end of Foula, and grows on two of the many islands in the West Loch, Hildasay, near Scalloway, where it was planted quite a long time ago.

Listera cordata. *Lesser Twayblade*. Widely but sparingly distributed throughout the major divisions of Mainland as well as in Unst, Yell, and Fetlar, and several of the smaller islands including Fair Isle and Foula. Damp mossy, moory or heathy, places on the hills, usually among or even under short heather bushes where it can easily be overlooked; rarely in the open in fine short heath.

Hammarbya paludosa. *Bog Orchid*. A very rare and inconspicuous orchid of wet sphagnous places in Yell and Papa Stour. There are five sites in Yell: by the W side of Bena Water, E of Kirk Loch (1961), fifty-five plants (2003, P. V. Harvey, A. Lockton, and S. Whild); on the sides of a small stream running to the coast, N of Bena Water, 107 flowering plants (2003, S. Duffield); by a small pool S of the head of the Burn of East Mires, SE of the Lochs of Lumbister, eight flowering spikes (1986); head of the Burn of Floga, North Sandwick, sixty-two plants, half of these in flower (1991); by the S side of the extreme E end of the Loch of Colvister, about twenty plants (1997). In Papa Stour a single plant was found by the tiny burn running into the Dutch Loch in 1980; there has been no serious attempt to refind it since that time.

[**Pseudorchis albida**. *Small-white Orchid*. Native, believed to be extinct. There are two records, one from Unst in 1806 by C. Fothergill, the other from Bressay before 1845 by T. Edmondston. Fothergill gives its habitat as moist pastures and meadows, a remark intended to cover three sites in Orkney as well as the one in Unst. Despite the fact that some of these recorders' unconfirmed records of other plants are open to doubt, there is little reason to query the above records, even in the absence of specimens. In Britain it is very much a Scottish plant, and extends to Orkney. It also occurs in Faeroe; furthermore, it bears no resemblance to any other Shetland orchid.]

Gymnadenia conopsea subsp. **borealis**. *Fragrant Orchid*. Very rare and after 1924 known only from dry serpentinite pastures N of Baltasound, Unst, where most plants are located within or just outside the Keen of Hamar National Nature Reserve/Site of Special Scientific Interest. In 1911 it was recorded from the district about Baltasound, almost certainly from the general area just described. Its numbers here vary markedly from year to year; in 2003—a favourable year for it—107 flowering plants were noted (W. Dickson, P. V. Harvey, and M. Maher). Two sites, each outside the general area but in the same serpentinite habitat, were found in 1962: near Halligarth (no later record), and near Hagdale (two colonies in 1984). Four old and unconfirmed records exist: near North Roe, Northmavine, 'dry places among heath' (before 1845); Clibberswick, Unst, a serpentinite area (*c.*1920); Skaw, Unst (1921); and Boddam, South Mainland (1924), presumably on heathy ground in both places.

Coeloglossum viride. *Frog Orchid*. A widely scattered but occasional species of short dry coastal pastures, often in sandy areas or cliff-top sward; also in short rocky turf on the limestone bands of Central Mainland, the serpentinite pastures and associated fellfield of Unst and Fetlar, and (rarely) on the granite fellfield about the Ronas Hill summit, Northmavine. Frog orchid is usually seen in small numbers, but in 1994 WS found it to be frequent on the Isle of Fethaland, Northmavine.

× **Dactyloglossum mixtum** (*Coeloglossum viride* × *Dactylorhiza fuchsii*). Very rare, and the first bigeneric orchid hybrid to be recorded from Shetland. Two plants fairly close together were observed (and photographed) in 2006 at Skeo Taing, Unst (HP642084), on the seaward side of the right angle bend in the road, by P. F. Hunt and R. Laurence. Its two parents, along with *Dactylorhiza purpurella*, grew nearby.

Dactylorhiza fuchsii. *Common Spotted-orchid.* Very rare—known only from four areas—but not uncommon and even frequent in parts of some of these places. Rich marshy or damp ground by lochs and burns on sandy soil where it is often among tall vegetation (notably horsetails); also, but less frequently, in fairly dry or wet pastures on or near sand. Its main area is in the S of South Mainland, from Sumburgh through the Links of Quendale to Spiggie and Scousburgh (including the Lochs of Brow and Spiggie); also in the Clevigarth, Mails, and Boddam areas to the E, and in the Troswick, Clumlie, and Braefield districts to the north. In some of these places it is very likely that agricultural and other activities have led to the destruction of some sites. A second area for this orchid in this division is found in the low marshy ground at the S end of Cunningsburgh. The remaining two stations are far to the N, in Unst; behind Sand Wick; and between Ordale and Skeo Taing (1989, R. C. Palmer).

D. × venusta (*D. fuchsii* × *D. purpurella*). This hybrid has been recorded, usually with the parents, in several places in the S of South Mainland, and in Cunningsburgh in the same division; also at the Burn of Burrafirth, Unst (2006, BSBI field meeting).

D. maculata. *Heath Spottted-orchid.* The commonest and most widespread of all of our orchids. A characteristic species of moist heathy pastures and moors, and absent only from the richest of soils. It is found in many places throughout the county.

D. × formosa (*D. maculata* × *purpurella*). A not infrequent, often tall and showy, hybrid between our two commonest orchids, and usually seen with or near one or both parents. Although it is usually found where the habitats of the parents are not far apart, it much prefers the marshier conditions favoured by *D. purpurella*.

D. incarnata. *Early Marsh-orchid.* Frequency, distribution, and habitat details are given under each of the three subspecies which grow in Shetland. Subsp. **incarnata**. Rich marshy pastures, base-rich flushes, and damp dune-pastures on limestone, sandstone, and serpentinite, etc. Local to frequent in South Mainland; frequent in Central Mainland; rare in Fair Isle, West and North Mainland, Unst and Fetlar; and unrecorded from Yell, Whalsay, and Bressay (apart from an old record from the Isle of Noss). Included in the foregoing are new records from Unst: Loch of Bordastubble (1989, M. G. Pennington), scattered over a wide area about the loch, and more generally from the Snarravoe area N to the Loch of Vigga (1992, WS); also from the Keen of Hamar National Nature Reserve/Site of Special Scientific Interest (2006, BSBI field meeting). There is also one new Fetlar record: Funzie mire (1991, M. Tickner). Subsp. **coccinea**. A local form with striking crimson-pink flowers in damp, or marshy, sandy pastures in the S of South Mainland has been noted at Sumburgh (by the airport runways, 1963, and near the Sumburgh Hotel, 1999 (both WS)); Links of Quendale, including (2003, WS) a particularly fine colony 600 m SW of the Loch of Huesbreck); Loch of Clevigarth area (2005, WS); and at St Ninian's Isle, two plants (1996, R. C. Palmer). Although presently treated as a subspecies, this distinction does not apply, at least in Shetland, where every gradation of flower colour from the normal flesh-pink of subsp. *incarnata* to the crimson-pink of subsp. *coccinea* can be found in most areas containing the latter subspecies, often in the same population.   Subsp. **pulchella**. Very rare. This purplish-pink subspecies is known with certainty only from three places on dry heathy serpentinite pastures to the N of Baltasound, Unst, where it seems always to be in small quantity: western slopes of the Keen of Hamar; a few specimens on the SE side of Little Heog, above the road (1991, L. A. Inkster and WS); groups and single plants below the road (above Hagdale), and one plant behind the nearby garage (same date and recorders).   There is an old (1952) and somewhat qualified record from Burrafirth, Unst, in need of confirmation.

D. × **latirella** (*D. incarnata* × *D. purpurella*).   Very rare.   Found sparingly in 1993, with both parents, in marshy ground by the SW side of the Loch of Tingwall, Central Mainland; also a single plant farther N on the W side of the loch, W of the Holm of Setter (1996, L. A. Inkster).

D. **purpurella**. *Northern Marsh-orchid.*   Our commonest orchid after *D. maculata*.   A frequent plant of marshy places and damp pastures (notably by lochs and burns in low-lying districts on the richer soils); also in quite dry ground as on the serpentinite debris of Unst, and (somewhat surprisingly) on dry grassy or gravelly roadsides.

Orchis mascula. *Early-purple Orchid.*   Very rare.   Known only from dry serpentinite fellfield and associated turfy pasture in Unst and Fetlar (including a few plants on greenstone on Huney, off Unst), and in rocky limestone pastures on the White Ness peninsula, Central Mainland.   In Unst it occurs from Little Heog to the Wick of Hagdale area, and on the western slopes of the Keen of Hamar National Nature Reserve/Site of Special Scientific Interest, where, in the last-named area, M. G. Pennington in 1990 counted 1,219 flowering plants.   In nearby Fetlar it is found in the Hill of Mongirsdale and Gudna Lee areas, and NW to the East Neap cliff-tops (WS). At Whiteness, scattered plants have been seen from W of Easthouse to NW of Hoove, and in 1992 L. A. Inkster saw hundreds in flower not far from Brekk, in the same area.

REFERENCES

ANDERSON, L. F. (2003). 'Tussock Grass', in Goddard, P., *Island of Bressay, Shetland. Community Biodiversity Action Plan*: 24.
>http://www.livingshetland.org.uk/documents/BressayCommunityBAP.pdf<
Last accessed 18 February 2011.

BEEBY, W. H. (1887–1909). [Numerous articles and notes. See part 4.]

BRUMMITT, R. K., KENT, D. H., LUSBY, P. S. & PALMER, R. C. (1987). 'The history and nomenclature of Thomas Edmondston's endemic Shetland *Cerastium*'. *Watsonia*, **16**: 291–297.

BRYSTING, A. K. (2008). 'The arctic mouse-ear in Scotland—and why it is not arctic'. *Plant Ecology & Diversity*, **1**: 321–327.

CARRICK, A., GRIGOR-TAYLOR, F., JOHNSON, A., MOYLAN, E. & SIGRUN, R. (1997). *Shetland Botanical Survey 1997.* [Lerwick.] A report to the Shetland Islands Council.)

CRAIG-CHRISTIE, A. (1870). 'Notes of a Botanical Excursion to Shetland in 1868'. *Transactions and Proceedings of the Botanical Society* [of Edinburgh], **10**: 165–170.

DRUCE, G. C. (1922). 'Flora Zetlandica'. *Botanical Society and Exchange Club of the British Isles. Report for 1921,* **6** (supplement to pt. 3): 457–546. Also issued as a separately paginated offprint; Druce's own annotated copy, in the Fielding-Druce Herbarium in the Department of Plant Sciences, Oxford, is one of the latter.

—— (1925). 'Additions to the Flora Zetlandica'. *Botanical Society and Exchange Club of the British Isles. Report for 1924,* **7**: 628–657.

EDMONDSTON, A. (1809). *A view of the ancient and present state of the Zetland Islands,* **1**: 195–219. Edinburgh.

EDMONDSTON, T. (1845). *A Flora of Shetland.* Aberdeen.

[FLINT, P.]. (2008). 'Unusual little find'. *Shetland Times,* 22 August 2008: 37.

FOTHERGILL, C. (1806). 'A View of the Natural History of the Isles In a Series of Descriptive Catalogues of the Quadrupeds Birds Fish Insects Reptiles Shells Crustaceous Animals Zoophytes and Plants That have hitherto been discovered in the Orcades, Shetland, Fair Isle, and Fula'. MS in the Thomas Fisher Rare Book Library, University of Toronto, Canada. Copy in Shetland Museum and Archives, Hay's Dock, Lerwick, Shetland.

FRASER-JENKINS, C. R. (2007). 'The species and subspecies in the Dryopteris affinis group'. *Fern Gazette,* **18**: 1–26.

GEAR, S. (2008). *Flora of Foula.* Foula, Shetland. Foula Heritage. (The first comprehensive pictorial guide to the plants of one of our smaller islands.)

G[ORRIE], W. (1855). 'D[*actylis*] *caespitosa*', *in* Morton, J. C. (ed.), *A Cyclopedia of Agriculture, practical and scientific,* **1**: 601–602. Glasgow. (Contains much interesting information on tussac-grass.)

HENDERSON, D. M. & DICKSON, J. H. (1994). *A naturalist in the Highlands.* Edinburgh. (An account of James Robertson whose travels in Shetland in 1769 provides us with the earlist listing of the plants of our area.)

JERMY, C., SIMPSON, D., FOLEY, M. & PORTER, M. (2007). *Sedges of the British Isles.* B.S.B.I. Handbook, no. 1 (ed. 3). London.

MACPHERSON, P., DICKSON, J. H., ELLIS, R. G., KENT, D. H. & STACE, C. A. (1996). 'Plant status nomenclature'. *BSBI News,* no. 72: 13–16.

PALMER, R. C. & SCOTT, W. (1969). *A Check-list of the Flowering Plants and Ferns of the Shetland Islands.* Published by the authors.

PRESTON, C. D., PEARMAN, D. A. & DINES, T.A. (2002. *New atlas of the British & Irish flora.* Oxford.

RAEBURN, H. (1888). 'The Summer Birds of Shetland, with Notes on their Distribution, Nesting, and Numbers'. *Proceedings of the Royal Physical Society of Edinburgh,* **9**: 542–562.

—— (1891). 'The birds of Papa Stour, with an account of the Lyra Skerry'. *The Zoologist,* **15** (third series): 126–135.

ROBERTSON, J. 1769. 'Flora and Fauna of Shetland'. Henderson & Dickson (1994) record that this MS was found in the Signet Library, Parliament Square, Edinburgh.

—— *c.*1771. 'Flora and Fauna of the Islands of Scotland'. Henderson & Dickson (1994) record that this MS was found in the Signet Library, Parliament Square, Edinburgh.

SCOTT, W., HARVEY, P., RIDDINGTON, R. & FISHER, M. (2002). *Rare plants of Shetland.* Lerwick.

SCOTT, W. & PALMER, R. [C.] (1987). *The flowering plants and ferns of the Shetland Islands.* Lerwick.

—— (1999). 'Flowering plants and ferns', *in* Johnston, J. L., *A Naturalist's Shetland*, 385–409. London.

SELL, P. & MURRELL, G. (2006). *Flora of Great Britain and Ireland*, **4**. Cambridge.

SHETLAND ISLANDS COUNCIL (2008). *Shetland in Statistics*, no. 35: 22.

SPENCE, D. (1979). *Shetland's living landscape.* Sandwick, Shetland.

SPENCE, M. (1914). *Flora Orcadensis.* Kirkwall.

STACE, C. [A.] (1997). *New flora of the British Isles*, ed. 2. Cambridge.

—— (2010). *New flora of the British Isles*, ed. 3. Cambridge.

STACE, C. A., ELLIS, R. G., KENT, D. H. & McCOSH, D. J. (2003). *Vice-county census catalogue of the vascular plants of Great Britain.* London.

TATE, R. (1866). 'Upon the flora of the Shetland Isles'. *Journal of Botany, British and foreign*, **4**: 2–15.

TYLER, G. (2003). 'Cape Pond Weed at the Vaadal Reservoir'. *Fair Isle Times*, 5 June 2003: [3].

WEST, W. (1912). 'Notes on the flora of Shetland, with some ecological observations'. *Journal of Botany, British and foreign*, **50**: 265–275, 297–306.

Index to Latin names in the checklist

Eriophorum 82
Erodium 39
Eryngium 40
Erysimum 21
Euphorbia 37
Euphrasia 53
Fagopyrum 16
Fallopia 16
Festuca 87
x Festulolium 88
Filipendula 28
Fragaria 29
Fraxinus 49
Fuchsia 36
Fumaria 10
Galeopsis 47
Galinsoga 76
Galium 57
Gaultheria 25
Gentianella 43
Geranium 38
Geum 29
Gladiolus 98
Glaucium 10
Glaux 27
Glechoma 47
Glyceria 91
Gnaphalium 72
Gymnadenia 99
Gymnocarpium 7
Hammarbya 99
Hebe 52
Hedera 40
Helianthus 76
Helichrysum 72
Helictotrichon 91
Heracleum 42
Hesperis 21
Hieracium 64
Hippuris 48
Holcus 92
Honckenya 13
Hordeum 95
Hosta 97

Humulus 11
Huperzia 4
Hyacinthoides 98
Hydrocotyle 40
Hymenophyllum 6
Hyoscyamus 44
Hypericum 18
Hypochaeris 60
Iberis 23
Ilex 37
Impatiens 39
Inula 72
Iris 98
Isoetes 5
Isolepis 84
Jasione 56
Juncus 80
Juniperus 8
Kickxia 50
Laburnum 34
Lactuca 60
Lamium 46
Lappula 46
Lapsana 59
Larix 8
Lathyrus 33
Lemna 79
Leontodon 60
Lepidium 23
Leucanthemum 74
Levisticum 42
Leymus 95
Ligusticum 42
Ligustrum 49
Lilium 97
Limnanthes 39
Linaria 50
Linum 37
Listera 99
Lithospermum 44
Littorella 49
Lobelia 56
Lobularia 22
Loiseleuria 25
Lolium 88

Lonicera 57
Lotus 32
Lunaria 22
Lupinus 34
Luzula 82
Lychnis 15
Lycopersicon 44
Lycopodium 4
Lysimachia 26
Malcolmia 21
Malus 31
Malva 18
Matricaria 74
Matthiola 21
Meconopsis 10
Medicago 33
Melampyrum 53
Melilotus 33
Mentha 47
Menyanthes 44
Mertensia 45
Mimulus 49
Minuartia 13
Molinia 96
Montia 12
Mycelis 60
Myosotis 45
Myriophyllum 35
Myrrhis 40
Narcissus 98
Nardus 87
Narthecium 97
Nymphaea 9
Nymphoides 44
Odontites 55
Ophioglossum 5
Orchis 101
Oreopteris 6
Ornithogalum 97
Osmunda 5
Oxalis 37
Oxyria 17
Panicum 96
Papaver 10
Parentucellia 55

Parnassia 28
Pastinaca 42
Pedicularis 55
Persicaria 15
Petasites 76
Peucedanum 42
Phacelia 44
Phalaris 93
Phegopteris 6
Phleum 94
Phragmites 96
Phyllitis 6
Picea 8
Picris 60
Pilosella 64
Pimpinella 40
Pinguicula 56
Pisum 33
Plagiobothrys 45
Plantago 49
Poa 89
Polemonium 44
Polygala 37
Polygonum 16
Polypodium 6
Polystichum 7
Populus 19
Potamogeton 76
Potentilla 29
Primula 26
Prunella 47
Pseudorchis 99
Pteridium 6
Puccinellia 89
Pulmonaria 44
Pyrola 26
Radiola 37
Ranunculus 9
Raphanus 24
Rapistrum 24
Rheum 17
Rhinanthus 55
Ribes 27
Rorippa 21
Rosa 31
Rubus 28

Rumex 17
Ruppia 79
Sagina 14
Salicornia 12
Salix 19
Sambucus 57
Saussurea 58
Saxifraga 28
Scandix 40
Schoenoplectus 83
Schoenus 84
Scilla 98
Sedum 27
Selaginella 5
Senecio 75
Setaria 96
Sherardia 57
Sidalcea 18
Silene 15
Sinapis 24
Sisymbrium 20
Solanum 44
Soleirolia 11
Solidago 72
Sonchus 60
Sorbus 31
Sparganium 97
Spergula 14
Spergularia 15
Spiraea 28
Stachys 46
Stellaria 13
Suaeda 12
Subularia 23
Succisa 58
Symphoricarpos 57
Symphytum 44
Tanacetum 73
Taraxacum 60
Thalictrum 10
Thermopsis 34
Thlaspi 23
Thymus 47
Trichophorum 82
Trientalis 26
Trifolium 33

Triglochin 76
Tripleurospermum 74
Trisetum 92
Triticum 96
Trollius 9
Tropaeolum 39
Tussilago 76
Typha 97
Ulex 34
Ulmus 11
Urtica 11
Utricularia 56
Vaccinium 25
Valerianella 57
Veronica 51
Vicia 32
Viola 18
Vulpia 89
Zannichellia 79
Zostera 79

Index to English names in the checklist

Part 2

A key to the species of *Hieracium* (hawkweeds) of the Shetland Islands

In fine, the groups into which the genus [*Hieracium*] is divided do not stand apart from each other by any such clear line of demarcation as the definitions of the systematists would lead us to suppose. There is no absolute conformity to type, but only approximation. Nature here, as elsewhere, refuses to be fitted into any exact compartments, or to submit to be limited by rigid theoretic formulae. W. R. Linton, *An account of the British Hieracia* (1905).

Introduction. Twenty-six species of *Hieracium* are on record for Shetland. Of these no fewer than eighteen are considered to be endemic to the area, including *H. hethlandiae,* exterminated in its only certainly known native station, and now seen in cultivation only, apart from one or two sites in Shetland where it has been planted. *H. maritimum*, which occurs elsewhere in Britain, is also believed to be extinct; it was first and last seen, in 1902, on a holm in a West Mainland loch. Because total extinction is very difficult to prove, these species appear in this key in the hope that they may yet be found again somewhere in Shetland. (Another hawkweed may some day be added to the local list, should our *H. sparsifolium* be found to contain two taxa as was first suggested in 1988 by J. Bevan. However, this hawkweed occurs elsewhere in the British Isles, and in Faeroe, Iceland, and Norway. This makes it inadvisable to describe a new taxon from a small area within the wider limits of *H. sparsifolium* without knowing the extent of its variability in those other areas. Such a study (which, ideally, should include the cultivation of material from many areas) may not be undertaken for a long time, and may only, after all, prove that *H. sparsifolium* is not satisfactorily divisible into smaller units. For the time being, this possible new taxon, to which the writer has assigned the label 'Taxon G', is incorporated within *H. sparsifolium* in this key without distinction).

General. Sectional features are given only where relevant to Shetland. All notes have been made on material growing naturally in Shetland, or on plants grown in Scalloway in pots (not in good garden soil where growth can be untypical), and on the numerous specimens from Shetland, many of which were gathered well over a century ago by W. H. Beeby, and preserved in various herbaria throughout the country.

Species. The twenty-six taxa (including the possibly extinct *H. maritimum)* are spread over six sections of the genus as follows. Nomenclature, and sequence of the sections follow Sell & Murrell (2006), but the species are listed alphabetically within each section. Species believed to be endemic to Shetland, which includes all the local members of Section Alpestria, are indicated by an obelus. (1) Section **Foliosa** (Fr.)

Arv.-Touv. One species. *H. maritimum* (F. Hanb.) F. Hanb. (2) Section **Tridentata** (Fr.) Arv.-Touv. Three species. *H. gothicoides* Pugsley, *H. lissolepium* Roffey, *H. sparsifolium* Lindeb. (3) Section **Alpestria** (Fr.) Arv.-Touv. Sixteen species. †*H. amaurostictum* Walter Scott & R. C. Palmer, †*H. attenuatifolium* P. D. Sell & C. West, †*H. australius* (Beeby) Pugsley, †*H. breve* Beeby, †*H. difficile* P. D. Sell & C. West, †*H. dilectum* P. D. Sell & C. West, †*H. gratum* P. D. Sell & C. West, †*H. hethlandiae* (F. Hanb.) Pugsley,†*H. klingrahoolense* Walter Scott & R. C. Palmer, †*H. northroense* Pugsley, †*H. praethulense* Pugsley, †*H. pugsleyi* P. D. Sell & C. West, †*H. spenceanum* Walter Scott & R. C. Palmer, †*H. subtruncatum* Beeby, †*H. vinicaule* P. D. Sell & C. West, †*H. zetlandicum* Beeby. (4) Section **Oreadea** (Fr.) Dahlst. Three species. *H. beebyanum* Pugsley, †*H. scottii* P. D. Sell, *H. subscoticum* P. D. Sell. (5) Section **Stelligera** Zahn. Two species. *H. argenteum* Fr., *H. caledonicum* F. Hanb. (6) Section **Hieracium**. One species. †*H. ronasii* P. D. Sell.

Leaves. Two kinds of leaves are found in hawkweeds. *Basal leaves* occur at soil level and often form a distinct rosette at flowering time. Where this happens the number of stem leaves may be as low as one (occasionally none), and does not exceed seven in Shetland material. *Stem leaves* are those which occur on the stem from just above soil level up to and including the leaf below the lowest bract. Bracts are excluded even when they are leaf-like in appearance. In this key basal and stem leaves are collectively referred to as leaves. The shape of the *middle stem leaves* is most useful and often gives a species its characteristic aspect. They should be chosen from the middle of the stem or just below, from average, undamaged plants which are neither too small nor overly developed. It is important to note that, as far as this key is concerned, the number of stem leaves *includes* those which have withered or fallen away at flowering time. A tiny mark on the stem indicates their former presence. *Leaf length* is measured from the point of origin on the stem to the tip of the apex, unless petiole length only is specified. Stem leaf number is given for all taxa.

Style colour. The colour of the two recurved branches of the style has been found to be very reliable for each species, not only within a population, but also in widely separated colonies, and in cultivation. It should always be noted on fully exserted styles on freshly opened capitula. In herbarium material the drying process can affect style colour; however, dark styles retain their colour in the herbarium. Style colour in Shetland hawkweeds falls easily into three groups. (1) *Yellow.* Indistinguishable or nearly so from the colour of the ligules of the same capitulum (as in *H. argenteum*), including those species where the styles have a faint orange tinge (as in *H. subtruncatum*). (2) *Pale livid.* A pale olive-green which slightly discolours the capitulum (as in *H. gratum*). (3) *Dark livid.* This is an obviously darker shade of olive-green which markedly discolours the capitulum (as in *H. breve*).

Flowering time. Shetland hawkweeds vary from species with a distinct rosette and one stem leaf (sometimes none) to plants in which no basal leaves are produced at all, but with over twenty stem leaves. The hawkweed student will soon become aware that those species with very few stem leaves flower earlier than those in which the stem leaves are numerous: *H. ronasii,* for example, is in bloom approximately one

116

month before *H. klingrahoolense.* In Shetland the flowering period for the genus as a whole extends from the end of June to the end of August.

Couplets. Perhaps inevitably, some couplets will contain leads which are not entirely mutually exclusive. This may occur, for example, where the number of stem leaves overlaps slightly, or where the user's perception of the contrasted matter may be subjective, as in leaf colour or the extent to which leaf clasping is manifested. Where there is doubt the writer would advise the user to follow both leads at first. This should indicate fairly soon which one is to be followed. Additional, secondary aids to identification are given within brackets for some species at the point where they key out. These extra features, while often useful, should only be regarded as possible enhancements to an identification based initially on the primary leads.

1(a). Stem leaves (0–) 1 (–2); leaves with strong reddish- or purplish-brown blotches; phyllaries with numerous glandular hairs; styles dark livid. (Basal leaves forming a distinct rosette at flowering time.) (Section Hieracium.) .... **H. ronasii**
1(b). Stem leaves 0–22, but if 0 or 1 then combination of leaf markings, phyllaries and styles not as above. (All other sections.) **2**

2(a). Stem leaves (0–) 1–7; basal leaves normally forming a distinct rosette at flowering time. (Sections Alpestria (in part), Oreadea, and Stelligera.) **3**
2(b). Stem leaves (3–) 4–22; basal leaves either never produced, or (when produced) withered or withering at flowering time. (Sections Foliosa, Tridentata, and Alpestria (in part **10**

3(a). Phyllaries moderately to abundantly glandular, sparingly to moderately pilose; stem leaves half-clasping; diameter of capitula 25–40 mm **4**
3(b). Phyllaries sparingly to moderately (rarely abundantly) glandular, moderately to abundantly (rarely sparingly) pilose; stem leaves not obviously clasping; diameter of capitula 40–50 mm **6**

4(a). Styles dark livid. (Stem leaves 3–7.) .... **H. spenceanum**
4(b). Styles pale livid. (Stems leaves 2–5.) **5**

5(a). Leaves denticulate to sharply serrate with forward-pointing teeth, olive green, stem leaves 2–3. (Upper stem leaves often short, broad, and prominently toothed; buds narrow) .... **H. zetlandicum**
5(b). Leaves denticulate, never sharply serrate, pale green to yellowish-green; stem leaves 3–5 .... **H. gratum**

6(a). Leaves with purplish blotches. (Leaves often silvery, glaucous green; styles pure yellow; stem leaves 0–2.) .... **H. argenteum**
6(b). Leaves always green and unmarked.. **7**

7(a). Stem leaves 0–1 (–2), the lower (or only) often large, the other (when present) small and bract-like. (Basal leaves abruptly narrowed into a shortly cuneate or subtruncate base; leaves bluish- or yellowish-green, firm.) .... **H. caledonicum**
7(b). Stem leaves (1–) 2–6 .... **8**

8(a). Phyllaries abundantly glandular and sparingly pilose. (Capitula noticeably large, to 50 mm diameter; buds to 10 mm diameter when ligule tips first appear; basal leaves forming a distinct rosette at flowering time; stem leaves 3–5.) .... **H. scottii**
8(b). Phyllaries abundantly pilose and sparingly to moderately glandular. (Capitula smaller, to 40 mm diameter; buds 8–9 mm diameter when ligule tips first appear **9**

9(a). Stem leaves (1–) 2–3, in lower half of stem, occasionally prominently toothed; some or all of the basal leaves present at flowering time; leaves often slightly bristly above and also on the margins, pale to dark green, occasionally glaucous green. (Our most frequently encountered hawkweed in the north of West Mainland and the west of North Mainland. It is usually referred to as *H. orimeles*, but for the present is included under *H. beebyanum*.) .... **H. beebyanum**
9(b). Stem leaves 4–6, more or less evenly spaced, normally fairly prominently toothed; basal leaves withered or withering at flowering time; leaves softly hairy, often dusky green .... **H. subscoticum**

10(a). Most or all of the leaves tapering gradually to a narrow attachment to the stem, the middle more rounded or cuneate at the base; lowest leaves more or less sessile or petiolate, the petioles up to 50 mm or more long. (Leaves sometimes aggregated at the base.) (Section Tridentata.) **11**
10(b). All leaves half-clasping and broadly attached to the stem, commonly widest below the middle, abruptly rounded (sometimes widened) at the base; lowest leaves sessile or with petioles up to 35 mm long. (Sections Foliosa, and Alpestria (in part).) ..
.... **13**

11(a). Leaves heavily marked with reddish- or purplish-brown blotches or streaks (paler in shade), very rarely green and unmarked; stem leaves 4–11; phyllaries with numerous glandular hairs. (Lowest leaves petiolate, the petioles up to 50 mm or more long; styles yellow with a faint orange tinge; plant usually tall.) (Section Tridentata.)
.... **H. sparsifolium**
11(b). Leaves pale green and unmarked; stem leaves 4–18; phyllaries with few glandular hairs **12**

12(a). Stem leaves 4–8, relatively evenly spaced along the stem; styles yellow...........
.... **H. gothicoides**
12(b). Stem leaves 9–18, tending to be crowded towards base of stem; styles pale livid .... **H. lissolepium**

13(a). Phyllaries epilose; stem leaves 12–20, the lowest sessile. (Basal leaves not produced. It has been been found once in Shetland, on a freshwater holm in West Mainland, where it is may be extinct.) (Section Foliosa.) .... **H. maritimum**

13(b). Phyllaries very sparingly to abundantly pilose, never epilose; stem leaves 4–22, the lowest commonly with a usually short, winged petiole, rarely sessile to subpetiolate. (Leaves often tinged purple below.) (Section Alpestria (in part).) **14**

14(a). Leaves with reddish- or purplish-brown blotches or streaks, the markings often very pale; stem leaves 3–6 .... **H. amaurostictum**
14(b). Leaves green and unmarked; stem leaves 4–22 .... **15**

15(a). Stem leaves up to 2.5 times longer than broad; plant short and compact, typically 15–20 cm high. (Styles dark livid; leaves deep green; stem leaves 4–11.) .... **H. breve**
15(b). Stem leaves 2.5–8.0 times longer than broad; plant taller, 40–90 cm high. (Stem leaves 4–22.) **16**

16(a). Stem leaves 4–10; middle stem leaves 2.5–5.0 times longer than broad **17**
16(b). Stem leaves 7–22; middle stem leaves 3.0–8.0 times longer than broad **19**

17(a). Middle stem leaves 2.5–4.0 times longer than broad, not narrowed to a long acute apex; styles dark livid. (Stem leaves 5–10.) .... **H. australius**
17(b). Middle stem leaves 3.5–5.0 times longer than broad, narrowed to a long, acute apex; styles pale livid. (Stem leaves 4–9.) **18**

18(a). Middle stem leaves with broadly rounded, half-clasping base; leaves closely spaced; plant to 40 cm high. (Stem leavs 5–9. This hawkweed is thought to be extinct in the wild.) .... **H. hethlandiae**
18(b). Middle stem leaves with narrowly rounded, half-clasping base; leaves widely spaced; plant to 70 cm high. (Stem leaves 4–8.) .... **H. difficile**

19(a). Styles yellow; middle stem leaves 3.0–5.0 times longer than broad. (Stem leaves 7–13.) **20**
19(b). Styles very pale livid or dark livid; middle stem leaves 3.5–8.0 times longer than broad. (Stem leaves 7–22.) **23**

20(a). Middle stem leaves with broadly rounded, half-clasping base (sometimes broadest at the base); phyllaries with numerous pilose hairs. (Stem leaves 8–12; basal leaves not produced.) .... **H. pugsleyi**
20(b). Middle stem leaves with narrowly rounded, half-clasping base; phyllaries with few pilose hairs. (Stem leaves 7–13.) **21**

21(a). Middle stem leaves often broadest above the middle; capitula large, to 50 mm diameter; phyllaries with abundant, markedly unequal. glandular hairs. (Stem leaves 8–10; basal leaves not produced.) .... **H. attenuatifolium**
21(b). Middle stem leaves elliptical (broadest at the middle and tapering evenly to both ends); capitula smaller, to 40 mm diameter; phyllaries with very few, short, glandular hairs (if many then stems and leaves nearly glabrous). (Stem leaves 7–13.) **22**

22(a). Stems often densely pilose-hairy, especially below; upper surface of leaves moderately pilose-hairy; phyllaries with very few glandular hairs. (Leaves dusky-olive-green, often markedly erect; styles yellow with a faint orange tinge; stems often zig-zag; stem leaves 7–13; basal leaves not produced.) .... **H. subtruncatum**
22(b). Stems thinly pilose-hairy or nearly glabrous; upper surface of leaves with few pilose hairs or nearly glabrous; phyllaries with few to many glandular hairs. (Leaves bright green; a few small basal leaves sometimes persisting to flowering time; stem leaves 7–13.) .... **H. praethulense**

23(a). Stem and leaves with very sparse hair-clothing of any type; middle stem leaves gradually narrowed to a broadly rounded, half-clasping base, the lower margins often subparallel; lowest stem leaves narrowed to a short, broadly or narrowly winged, half-clasping base, obscuring the distinction between petiole and blade; leaves subentire to minutely denticulate. (A few small basal leaves sometimes persisting to flowering time; stem leaves 10–22.) .... **H. klingrahoolense**
23(b). Stem and leaves with sparse to numerous pilose and/or floccose hairs; middle stem leaves with a narrow, half-clasping base; lowest stem leaves subpetiolate or with winged petioles up to 35 mm long; leaves denticulate to dentate. (Stem leaves 7–17.) **24**

24(a). Styles very pale livid; middle stem leaves noticeably drawn out to a long acute apex. (Middle stem leaves 4.5–5.5 times longer than broad; stem leaves 7–15; basal leaves not produced.) .... **H. dilectum**
24(b). Styles dark livid; middle stem leaves not noticeably drawn out to a long acute apex. (Middle stem leaves 3.5–8.0 times longer than broad; stem leaves 9–17.) **25**

25(a). Leaves closely denticulate; middle stem leaves to 65 mm long; phyllaries with few glandular hairs; plant 27–45 (–60) cm high. (Middle stem leaves 3.5–6.5 times longer than broad; a few small basal leaves rarely persisting to flowering time; stem leaves 9–13.) .... **H. northroense**
25(b). Leaves subentire to remotely denticulate, rarely strongly dentate; middle stem leaves to 85 mm long; phyllaries with numerous glandular hairs; plant to 70 cm high. (Middle stem leaves 4.0–8.0 times longer than broad; stem leaves 9–17; basal leaves not produced.) .... **H. vinicaule**

REFERENCE

SELL, P. & MURRELL, G. (2006). *Hieracium* L., in *Flora of Great Britain and Ireland*, **4**: 216–421. Cambridge.

Part 3

Images of herbarium material of the endemic species of *Taraxacum, Pilosella,* and *Hieracium* of the Shetland Islands

Plate 1. Taraxacum geirhildae (Shetland Dandelion)

Plate 2. T. serpenticola (Serpentine Dandelion)

Plate 3. T. hirsutissimum (Hairy Dandelion)

Plate 4. Pilosella flagellaris subsp. bicapitata (Shetland Mouse-ear-hawkweed)

Plate 5. Hieracium vinicaule (Wine-stemmed Hawkweed)

Plate 6. H. northroense (North Roe Hawkweed)

Plate 7. H. klingrahoolense (Klingrahool Hawkweed)

Plate 8. H. subtruncatum (Mainland Hawkweed)

Plate 9. H. dilectum (Purple-tinted Hawkweed)

Plate 10. H. pugsleyi (Pugsley's Hawkweed)

Plate 11. H. spenceanum (Spence's Hawkweed)

Plate 12. H. attenuatifolium (Laxo Burn Hawkweed)

Plate 13. H. hethlandiae (Cliva Hill Hawkweed)

Plate 14. H. praethulense (Thule Hawkweed)

Plate 15. H. australius (Unst Hawkweed)

Plate 16. H. difficile (Okraquoy Hawkweed)

Plate 17. H. amaurostictum (Semblister Hawkweed)

Plate 18. H. gratum (Handsome Hawkweed)

Plate 19. H. breve (Rare Hawkweed)

Plate 20. H. zetlandicum (Shetland Hawkweed)

Plate 21. H. scottii (Scott's Hawkweed)

Plate 22. H. ronasii (Ronas Voe Hawkweed)

(Each image is shown at about forty-four per cent of its original size)

SHETLAND ISLANDS

EX HB. W. SCOTT

Taraxacum geirhildae (Beeby)
R. C. Palmer & W. Scott

Grassy ledges and crevices on steep
west-facing rock-mass above east
side of Lang Clodie Loch, N. Roe.
Northmavine, Shetland (HU 3187).
7 June 1980.
Coll. W. Scott
Det. A. J. Richards.

No. 2692
[sheet 1 of 5]

Plate 1. **Taraxacum geirhildae** (Shetland Dandelion), p. 61

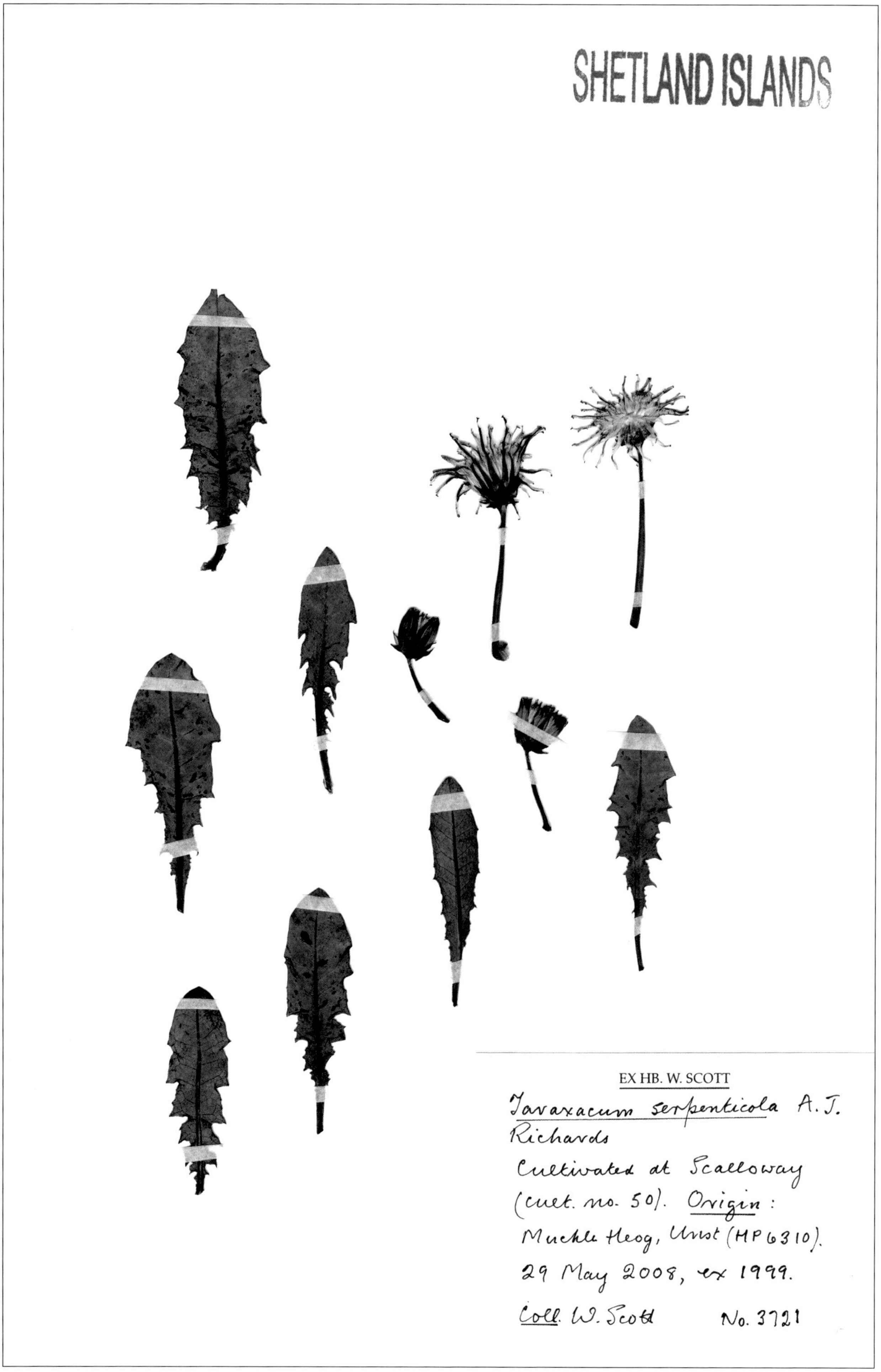
SHETLAND ISLANDS
EX HB. W. SCOTT
Taraxacum serpenticola A.J. Richards
Cultivated at Scalloway (cult. no. 50). Origin: Muckle Heog, Unst (HP 6310).
29 May 2008, ex 1999.
Coll. W. Scott No. 3721

Plate 2. **Taraxacum serpenticola** (Serpentine Dandelion), p. 61

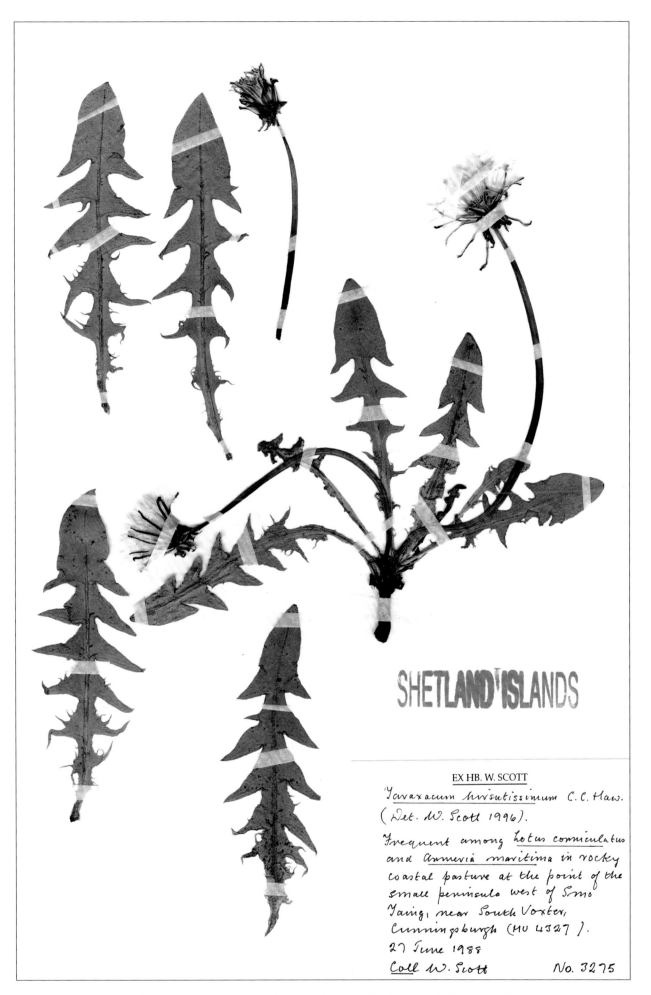

SHETLAND ISLANDS

Plate 3. **Taraxacum hirsutissimum** (Hairy Dandelion), p. 62

SHETLAND ISLANDS

EX HB. W. SCOTT

Pilosella flagellaris (Willd.) Sell &
West subsp. *bicapitata* Sell & West

Frequent in dry, rocky,
heathy pasture, West
Burrafirth, West Mainland
(HU 25).
3 July 1967
Coll. W. Scott
No. 1744
(Shown at 1967 BSBI Exhib. Meeting.)

(Sheet 1 of 3)

Plate 4. **Pilosella flagellaris** subsp. **bicapitata** (Shetland Mouse-ear-hawkweed), p. 64

EX HB. W. SCOTT

Hieracium vinicaule Sell & West
(det. W. Scott).

Several patches on rocky sea-
banks (with *Luzula sylvatica*) on
the west side of Gluss Isle, below
Ramna Hill (HU 3678).
6 August 1992
Coll. L. A. Inkster & W. Scott
No. 3164

SHETLAND ISLANDS

Plate 5. **Hieracium vinicaule** (Wine-stemmed Hawkweed), p. 65

SHETLAND ISLANDS

EX HB. W. SCOTT

Hieracium northroense Pugsley
(Det. W. Scott, 1987).

Common on small rocky knoll
in meadow below Burravoe,
North Roe (HU 3789), with
H. zetlandicum.
24 August 1987
Coll. W. Scott   No. 3264
                    (Sheet 1 of 3)

Plate 6. **Hieracium northroense** (North Roe Hawkweed), p. 66

EX HB. W. SCOTT

Hieracium

Steep rocks by pool, lower
end of the Burn of Skelladale,
near Brae (HU 3667).

15 August 1989
Coll. W. Scott
No. 3286
(Sheet 1 of 4)

Hieracium klingrahoolense
Walter Scott & R.C. Palmer

Det. W. Scott 25/9/2006

Formerly referred to
as 'Taxon C'.

Det. W. Scott 2002

Plate 7. **Hieracium klingrahoolense** (Klingrahool Hawkweed), p. 66

SHETLAND ISLANDS

Hieracium subtruncatum Beeby
(Det. W. Scott 1993).

Frequent on steep rocky sea-bank,
near and among honeysuckle,
west side of Stromness Voe, c.
300m. north of Jackville (HU 3743).
9 August 1993.
Coll. L. Farrell, L.A. Inkster,
W. Scott and D. Smith
                                No. 3259

Plate 8. **Hieracium subtruncatum** (Mainland Hawkweed), p. 66

SHETLAND ISLANDS

Hieracium dilectum Sell & West

Ravine at the lower end of the
Burn of Quoys, South
Nesting (HU 4454).

3 August 1964
Coll. W. Scott   No. 135

Hieracium *dilectum* Sell & West

Determinavit   1964   P. D. Sell
                                C. West

Plate 9. **Hieracium dilectum** (Purple-tinted Hawkweed), p. 66

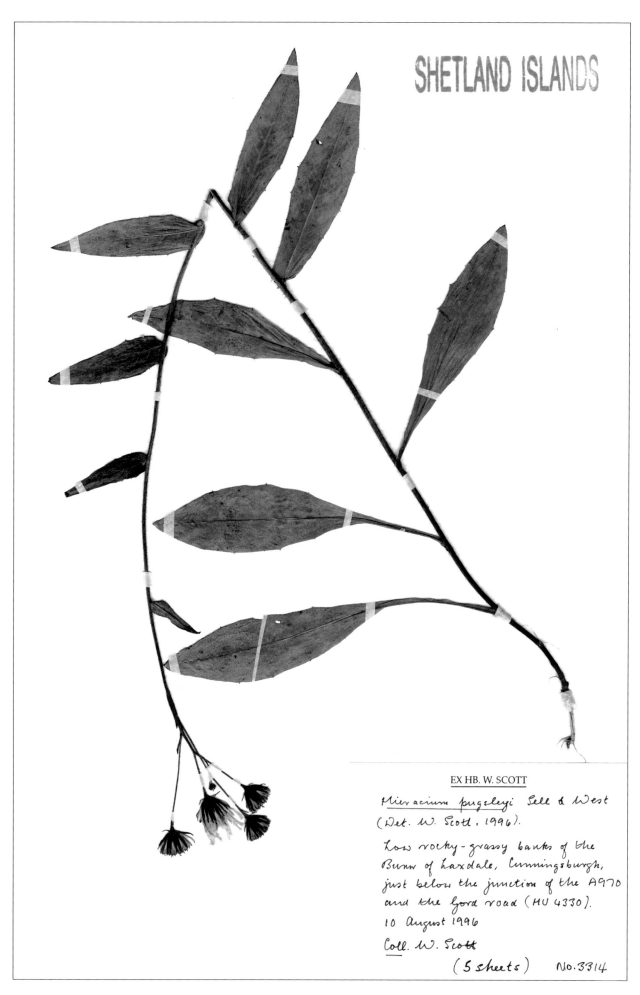

EX HB. W. SCOTT

Hieracium pugsleyi Sell & West
(Det. W. Scott, 1996).

Low rocky-grassy banks of the
Burn of Laxdale, Cunningsburgh,
just below the junction of the A970
and the Gord road (HU 4330).
10 August 1996
Coll. W. Scott
    (5 sheets)   No. 3314

Plate 10. **Hieracium pugsleyi** (Pugsley's Hawkweed), p. 66

SHETLAND ISLANDS

EX HB. W. SCOTT

Hieracium spenceanum W. Scott
& R.C. Palmer

Common in pastures and on sea-
banks from near Burravoe to
the head of West Burra Firth,
West Mainland, Shetland (HU 2556).

14 July 1963

coll. W. Scott   No. 1318

Det. W. Scott, 1995.

Hieracium attenuatifolium Sell & West

Determinavit   1963   P. D. Sell
                           C. West

Plate 11. **Hieracium spenceanum** (Spence's Hawkweed), p. 67

SHETLAND ISLANDS

Hieracium attenuatifolium Sell & C. West

On rocks at the lower end of the Laxo Burn, east coast of Mainland, Shetland (HU4463).

July 1956.

Coll. W. Scott   No. 1222

Named *H. attenuatifolium* Sell and West by DR. C. WEST, who saw this specimen at Lerwick, 1/8/64

Plate 12. **Hieracium attenuatifolium** (Laxo Burn Hawkweed), p. 67

SHETLAND ISLANDS

EX HB. W. SCOTT

Hieracium hethlandiae (F.J. Hanb.) Pugsley
Cultivated at Scalloway (cult.
no. 4). Origin: steep roadside
banks, Cliva Hill, Delting
(HU 3468), ex 1975/6.
8 August 2005
Coll. W. Scott   No. 3625

Plate 13. **Hieracium hethlandiae** (Cliva Hill Hawkweed), p. 67

SHETLAND ISLANDS

EX HB. W SCOTT

Hieracium praethulense Pugsley

Among rocks near The Kink,
north side of Ronas Voe
(HU 2883).
30 July 1967
Coll. W. Scott   No. 1728

Hieracium *praethulense* Pugsl.

Determinavit 1967.   P. D. Sell
                             C. West

Plate 14. **Hieracium praethulense** (Thule Hawkweed), p. 68

Hieracium australius (Beeby)
Pugsley
Cultivated at Scalloway (cult.
no. 117). Origin: Burrafirth
area, Unst (HP 61), ex 1988.
31 July 2005
Coll. W. Scott   No. 3623

SHETLAND ISLANDS

Plate 15. **Hieracium australius** (Unst Hawkweed), p. 68

SHETLAND ISLANDS

Hieracium _difficile_ Sell + West

Determinavit   19⁞7   P. D. Sell
                        C. West

EX HB. W. SCOTT

Hieracium difficile Sell & West

Banks of stream above Bay
of Okraquoy, south of
Fladdabister (HU 4331).

4 August 1967
Coll. W. Scott   No. 1734

Plate 16. **Hieracium difficile** (Okraquoy Hawkweed), p. 68

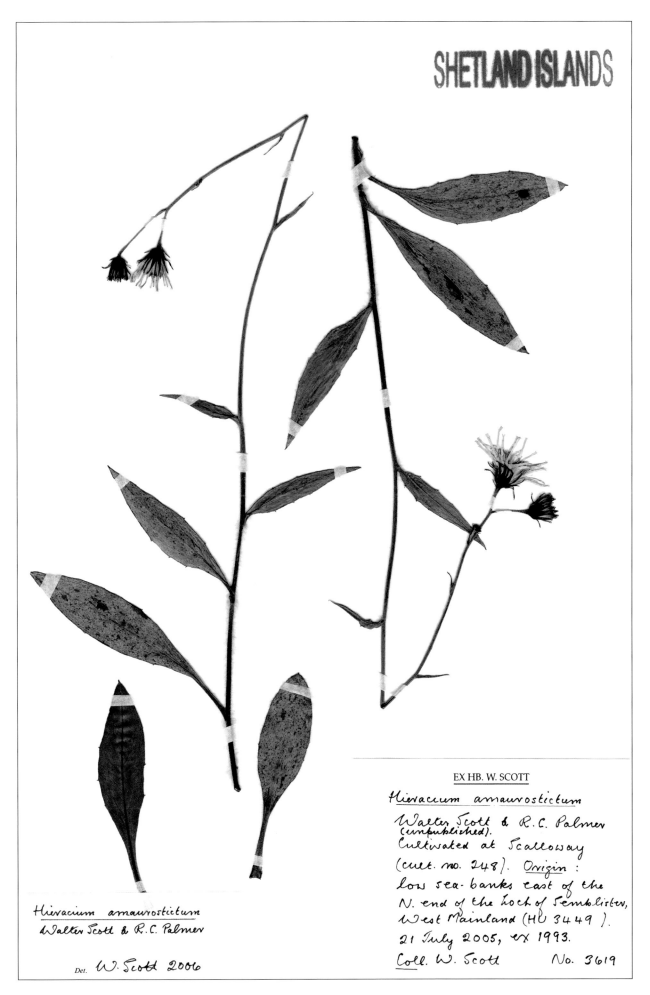

EX HB. W. SCOTT

Hieracium amaurostictum

Walter Scott & R.C. Palmer
(unpublished).
Cultivated at Scalloway
(cult. no. 248). Origin:
low sea-banks east of the
N. end of the Loch of Semblister,
West Mainland (HU 3449).
21 July 2005, ex 1993.
Coll. W. Scott   No. 3619

Hieracium amaurostictum
Walter Scott & R.C. Palmer

Det. W. Scott 2006

Plate 17. **Hieracium amaurostictum** (Semblister Hawkweed), p. 68

SHETLAND ISLANDS

Hieracium gratum Sell & West
(Det. W. Scott 1993).

Rocks high up at the west
end of Longi Geo, west side of
Whale Firth, Yell (HU 4695).

5 August 1993

Coll. W. Scott   No. 3261

Plate 18. **Hieracium gratum** (Handsome Hawkweed), p. 69

EX HB. W. SCOTT

Hieracium breve Beeby

Among rocks west of Feal,
Ronas Voe (HU 3081).
21 July 1963

Coll. W. Scott   No. 1299

Hieracium *breve* Beeby

Determinavit   1963   P. D. Sell
                          C. West

Plate 19. **Hieracium breve** (Rare Hawkweed), p. 69

SHETLAND ISLANDS

EX HB. W. SCOTT

Hieracium zetlandicum Beeby

Heathy pasture, west side of
Snarra Ness, West Mainland
(HU 2357).

14 July 1963

Coll. W. Scott   No. 1295

Hieracium zetlandicum Beeby

Determinavit   1963,   P. D. Sell
                       C. West

Plate 20. **Hieracium zetlandicum** (Shetland Hawkweeed), p. 69

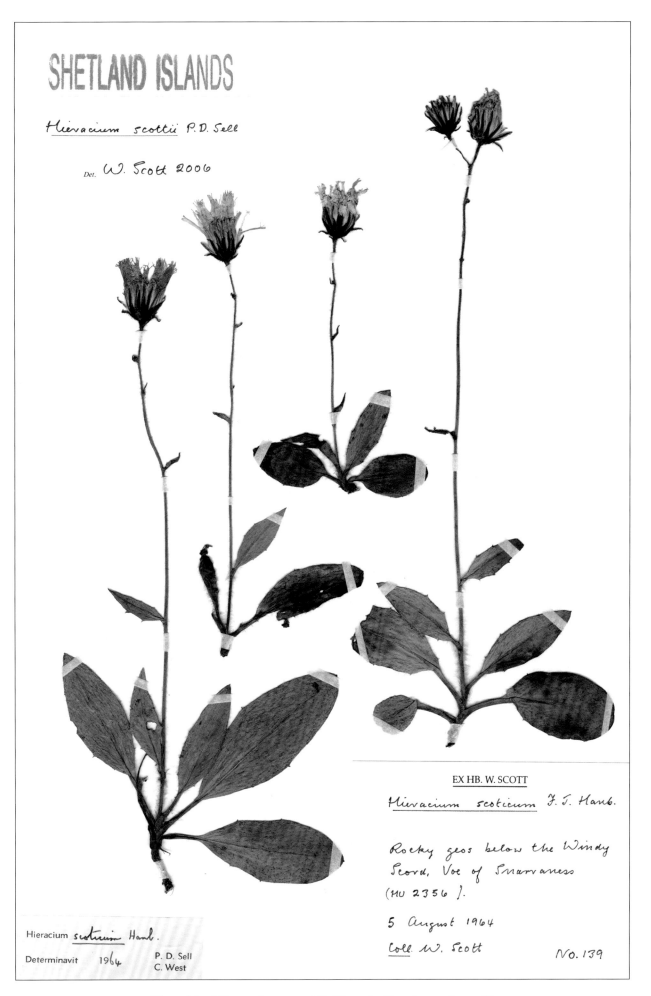

SHETLAND ISLANDS

Hieracium scottii P. D. Sell

Det. W. Scott 2006

EX HB. W. SCOTT

Hieracium scoticum F. J. Hanb.

Rocky geos below the Windy
Scord, Voe of Snarraness
(HU 2356).

5 August 1964
Coll. W. Scott   No. 139

Hieracium scoticum Hanb.

Determinavit   1964   P. D. Sell
                         C. West

Plate 21. **Hieracium scottii** (Scott's Hawkweed), p. 70

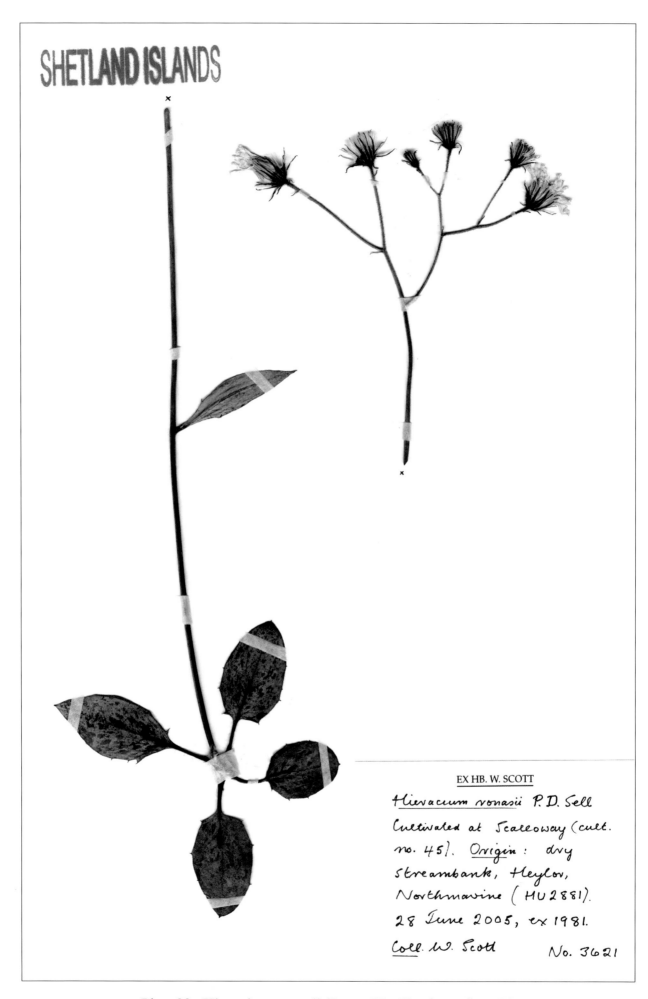

Hieracium ronasii P.D. Sell
Cultivated at Scalloway (cult.
no. 45). Origin: dry
streambank, Heylor,
Northmavine (HU2881).
28 June 2005, ex 1981.
Coll. W. Scott   No. 3621

Plate 22. **Hieracium ronasii** (Ronas Voe Hawkweed), p. 71

Part 4

A bibliography of the flowering plants, ferns, and fern allies of the Shetland Islands

This bibliography is based on the one in Scott & Palmer (1987) and updated to the present time. A few items in the earlier bibliography have been omitted because of their dwindling relevance or by their replacement by more modern and authoritative works. The present version contains many items which have a direct relevance to Shetland, from works of non-botanical interest in which a brief but nevertheless important or interesting observation on the vegetation may be found, to publications dealing directly or indirectly with at least one of the following aspects of our flora: distribution, taxonomy, habitats, and ecology. In addition to the foregoing there are also listings on the botany of some of our nearest neighbours, including the remainder of the British Isles to the south, and the Scandinavian countries of Faeroe and Iceland to the north-west. A perusal of these works is necessary in placing our island flora within the context of the floras of these other insular areas of north-west Europe. A few books devoted to horticulture in Shetland have been included to reflect the fact that a number of garden plants often appear outside cultivation, and where some have become familiar and well-established aliens. A short note is sometimes given where an item is of more than passing interest and appears in a publication which is not of any particular relevance to Shetland botany; a similar note is provided where the significance of an item is not evident from its title.

ADAIR, E. (1983). 'Flowers of Fetlar'. *Shetland Life*, no. 37: 28–31.

ALLEN, D. E. (1952). 'Cakile edentula (Bigel.) Hook. in Britain'. *Watsonia*, **2**: 282–283.

—— (1966). 'A list of infraspecific taxa of British phanerogams tested in cultivation'. *Watsonia*, **6**: 205–215.

ALLOTT, C. (1971). 'Report on the lowland vegetation of Foula. August 1970'. *Brathay Exploration Group. Field Studies Report*, no.11: 48–58.

ALMQUIST, E. (1921), introduced by G. C. Druce. 'Bursa pastoris Weber'. *Botanical Society and Exchange Club of the British Isles. Report for 1920*, **6** (supplement to pt.1): 179–207.

ANDERSON, J. (1879), ed. *A tour through the islands of Orkney and Schetland*, by G. Low. Kirkwall. (Many references to Shetland plants seen by Low in 1774.)

ANDERSON, J. (1967). 'The Botany of Shetland', *in* Cluness, A. T. (ed.), *The Shetland Book*, 141–158. [Lerwick.]

ANDERSON, L. F. (2003). 'Tussock Grass', in Goddard, P., *Island of Bressay, Shetland. Community Biodiversity Action Plan*: 24. (See over.)

>http://www.livingshetland.org.uk/documents/BressayCommunityBAP.pdf<
Last accessed 18 February 2011.

ANGUS, J. S. (1926). *Echoes from Klingrahool*, ed. 3. Lerwick. (A reference to wild roses at Catfirth, Central Mainland, on p. iii.)

[ANON.] (1711). 'Concerning the *Natural Advantages* of *Shetland*, for the Inhabitants, and their Usefulness to the Crown of Great-Britain', *in* Sibbald, R. (publ.), *The Description of the Isles of Orknay and Shetland*, 39–40. Edinburgh. (A few early references to Shetland plants (common scurvygrass, heather, etc.)

[ANON.] (1831). *A Walk in Shetland by two eccentrics*. Edinburgh. (An early mention of common nettle near Lerwick on p. 15.)

[ANON.] (1870). 'Appendix to Mr Craig-Christie's Paper on Plants collected in Shetland in 1868'. *Transactions of the Botanical Society* [of Edinburgh], **10**: 254–256.

[ANON.] (1907). 'Eighteenth Century Records of British Plants'. *Notes from the Royal Botanic Garden, Edinburgh*, **4**: 123–190.

[ANON.] (1949). 'Rare Plant in Mid Yell'. *Shetland News*, 10 November 1949: 2. (A note on herb-Robert.)

ARCHER, T. C. (1871). '*Note on the Changes produced in* Lotus corniculatus *by Cultivation*'. *Report of the Fortieth Meeting of the British Association for the Advancement of Science*, 109–110.

ATKINSON, G. C. 1833. 'An Excursion to the Shetland Islands. 1832'. Typescript in the Shetland Library, Lower Hillhead, Lerwick, Shetland. (One of Atkinson's companions was the ornithologist W. C. Hewitson.)

BABINGTON, C. C. (1847). '*Monograph of the British Atripliceae*'. *Transactions of the Botanical Society* [of Edinburgh], **1**: 1–17, plus two plates.

[BAIKIE, W. B.] (1847). *List of books and manuscripts relating to Orkney and Zetland*. Kirkwall.

BAILEY, J., HOLLINGSWORTH, P. & PRESTON, C. D. (1997). 'A cytologically confirmed population of *Potamogeton* × *gessnacensis* in Shetland'. *BSBI Scottish Newsletter*, no. 19: 6. (Repeated, almost verbatim, in *BSBI News*, no. 75: 62.)

BAILLIE, J. L. (1944). 'Charles Fothergill 1782–1840'. *Canadian Historical Review*, **25**: 376–396.

BAKER, H. G. (1947). 'Biological flora of the British Isles. *Melandrium* (Roehling em.) Fries'. *Journal of Ecology*, **35**: 271–292.

—— (1948). 'The ecotypes of *Melandrium dioicum* (L. emend.) Coss. & Germ.'. *New Phytologist*, **47**: 131–145.

—— (1956). '*Geranium purpureum* Vill. and *G. robertianum* L. in the British flora. II. *Geranium robertianum*'. *Watsonia*, **3**: 270–279.

BALL, D. F. & GOODIER, R. (1974). 'Ronas Hill, Shetland: a preliminary account of its ground pattern features resulting from the action of frost and wind', *in* Goodier, R. (ed.), *The natural environment of Shetland*, 89–106. Nature Conservancy Council, Edinburgh.

BALL, P. W. (1964). 'A revision of *Cakile* in Europe'. *Feddes Repertorium specierum novarum regni vegetabilis*, **69**: 35–40.

BANGERTER, E. B. & KENT, D. H. (1957). 'Veronica filiformis Sm. in the British Isles'. *Proceedings of the Botanical Society of the British Isles*, **2**: 197–217.

124

——— (1965). 'Additional notes on Veronica filiformis'. *Proceedings of the Botanical Society of the British Isles*, **6**: 113–118.

BARKHAM, J. P. (1971). 'A report on the upland vegetation of Foula – August 1968'. *Brathay Exploration Group. Field Studies Report*, no. 11: 25–47.

BARKHAM, J. P., GEAR, S., HAWKSWORTH, D. L. & MESSENGER, K. G. (1981). 'The flora of Foula'. *Brathay Exploration Group. Field Studies Report*, no 36. (Also described as 'Foula, Shetland. Volume 2. The flora of Foula'.)

BARRINGTON, R. M. (1890a). 'The great skua (*Lestris catarrhactes*) in Foula'. *The Zoologist*, **14** (third series): 297–301. (Barrington, an Irish botanist and ornithologist, noted dwarf willow, bog bilberry, and lesser twayblade on Foula.)

——— (1890b). 'Trientalis europaea in Foula'. *Journal of Botany, British and foreign,* **28**: 315–316.

BATEMAN, R. M. & DENHOLM, I. (1983). 'A reappraisal of the British and Irish dactylorchids, 1. The tetraploid marsh-orchids'. *Watsonia*, **14**: 347–376.

BEAVEN, E. S. (1947). *Barley. Fifty Years of Observation and Experiment.* London.

BEEBY, W. H. (1887a). 'On the flora of Shetland'. *Scottish Naturalist*, **3** (new series; **9** from start): 20–32.

——— (1887b). 'New British Plant'. *Scottish Naturalist*, **3** (new series; **9** from start): 184. (A note of a plant which later proved to be a form of one of our common sedges.)

——— (1887c). 'On *Ranunculus flammula*'. *Journal of Botany, British and foreign,* **25**: 370–372.

——— (1888). 'On the flora of Shetland'. *Scottish Naturalist*, **3** (new series; **9** from start): 209–218.

——— (1889). 'On the flora of Shetland'. *Scottish Naturalist*, **4** (new series; **10** from start): 32–39.

——— (1890a). 'On the flora of Shetland'. *Scottish Naturalist*, **4** (new series; **10** from start): 212–217.

——— (1890b). 'On Rumex propinquus J. E. Aresch., in Britain'. *Scottish Naturalist*, **4** (new series; **10** from start): 300.

——— (1891a). 'On the flora of Shetland'. *Scottish Naturalist*, **1** (third series): 25–30.

——— (1891b). 'Hieracium protractum Lindeb. in Britain'. *Journal of Botany, British and foreign,* **29**: 53.

——— (1891c). 'A new *Hieracium*'. *Journal of Botany, British and foreign,* **29**: 243–244. (A note on *Hieracium zetlandicum*.)

——— (1892). 'On the flora of Shetland'. *Annals of Scottish Natural History*, [**1**]: 51–55.

——— (1894a). 'Eleocharis acicularis'. *Journal of Botany, British and foreign,* **32**: 87.

——— (1894b), *in* Groves, J. (ed.), *Botanical Exchange Club of the British Isles. Report for 1893*. (References to Shetland plants on pp. 398, 399, 422, 424.)

——— (1895a), *in* Linton, W. R. (ed.), *Botanical Exchange Club of the British Isles. Report for 1894*. (A note on *Hieracium zetlandicum* on p. 456.)

—— (1895b). *Glyceria distans* var. *prostrata* Beeby. *Journal of Botany, British and foreign*, **33**: 315–316.

—— (1898). *Botanical Exchange Club of the British Isles. Report for 1896.* (References to Shetland plants on pp. 522, 525, 532.)

—— (1900), *in* Groves, J. (ed.), *Botanical Exchange Club of the British Isles. Report for 1898.* (An interesting note on the cultivation of Shetland mouse-ear on p. 568, and references to eyebrights on p. 583.)

—— (1901), *in* Linton, W. R. (ed.), *Botanical Exchange Club of the British Isles. Report for 1899.* (A note on the hawkweed now known as *Hieracium sparsifolium* on p. 608.)

—— (1907a). 'The Birch-tree in Shetland'. *Orkney and Shetland Miscellany*, **1**: 56, 91–92.

—— (1907b). 'On the flora of Shetland'. *Annals of Scottish Natural History*, [**16**]: 164–169, 233–239.

—— (1908a). 'On the flora of Shetland'. *Annals of Scottish Natural History*, [**17**]: 110–117.

—— (1908b). 'The scape of *Taraxacum*'. *Journal of Botany, British and foreign*, **46**: 120–123.

—— (1909a). 'On the flora of Shetland'. *Annals of Scottish Natural History*, [**18**]: 103–107.

—— (1909b). '*Euphrasia* and *Rhinanthus*'. *Annals of Scottish Natural History*, [**18**]: 234–237.

BENNETT, A. (1886). 'Additional records of plants from Scotland'. *Scottish Naturalist*, **2** (new series; **8** from start): 279–288, 309–319.

—— (1887). 'Additional records of Scotch plants for the year 1886'. *Scottish Naturalist*, **3** (new series; **9** from start): 56–66.

—— (1888). 'Additional records of Scottish plants for the year 1887'. *Scottish Naturalist*, **3** (new series; **9** from start): 247–261.

—— (1891). 'Record of Scottish plants for 1890, additional to "Topographical Botany". Ed. 2'. *Scottish Naturalist*, **1** (third series): 185–190.

—— (1894). 'Records of Scottish plants for 1893, additional to Watson's "Topographical Botany", 2nd Ed.'. *Annals of Scottish Natural History*, [**3**]: 158–164. (A note on grey field-speedwell on p. 164.)

—— (1897). 'Records of Scottish plants for 1896, additional to Watson's "Topographical Botany", 2nd Ed. (1883)'. *Annals of Scottish Natural History*, [**6**]: 246–248. (References to creeping soft-grass and small adder's-tongue.)

—— (1899). 'Records of Scottish plants for 1898, additional to Watson's "Topographical Botany", 2nd Ed. (1883)'. *Annals of Scottish Natural History*, [**8**]: 92–94. (References to butterbur (later found to be an error for winter heliotrope), and to common reed.)

—— (1901). 'Records of Scottish plants for 1900, additional to Watson's "Topographical Botany", 2nd Ed. (1883)'. *Annals of Scottish Natural History*, [**10**]: 100–102. (References to common stork's-bill and cornflower.)

—— (1905). 'Supplement to "Topographical Botany". Ed 2'. *Journal of Botany, British and foreign*, **43** (supplement). (Issued, from eight to sixteen pages per month, from March (no. 507) to November (no. 515).)

—— (1907). 'Forms of *Potamogeton* new to Britain'. *Journal of Botany, British and foreign*, **45**: 172–176.

—— (1909). 'Plants of the Faroe Isles not occurring in Great Britain, and others not occurring in Shetland, Orkney, Caithness, or the Outer Hebrides'. *Annals of Scottish Natural History*, [**18**]: 36–40.

BENNETT, A., SALMON, C. E. & MATTHEWS, J. R. (1929–1930). 'Second Supplement to Watson's "Topographical Botany" '. *Journal of Botany, British and foreign*, **67–68** (supplement).

BENNETT, K. D., BOREHAM, S., SHARP, M. J. & SWITSUR, V. R. (1992). 'Holocene history of environment, vegetation and human settlement on Catta Ness, Lunnasting, Shetland'. *Journal of Ecology*, **80**: 241–273.

BENOIT, P. M. & STACE, C. A. (1975). '*Galeopsis* L.', *in* Stace, C. A. (ed.), *Hybridization and the Flora of the British Isles*, 396–397. London.

BERRY, R. J. & JOHNSTON, J. L. (1980a). *The Natural History of Shetland*. London. (New Naturalist series, no. 64.)

—— (1980b). 'Lochs and burns', *in* Berry, R. J. & Johnston, J. L., *The Natural History of Shetland*, 118–127. London. (New Naturalist series, no. 64.)

BEVAN, J. (1988). 'A survey of Hieracia (Hawkweeds) in Shetland 1987'. (A report to the Nature Conservancy Council.)

—— (1990). 'A supplement to Hieracia (Hawkweed) sites in Shetland 1988'. (A report to the Nature Conservancy Council.)

BIRKS, H. J. B. & PEGLAR, S. M. (1979). 'Interglacial pollen spectra from Sel Ayre, Shetland'. *New Phytologist*, **83**: 559–575.

BIRKS, H. J. B. & RANSOM, M. E. (1969). 'An interglacial peat at Fugla Ness, Shetland'. *New Phytologist*, **68**: 777–796.

BIRNIE, R. V. & HULME, P. D. (1990). 'Overgrazing of peatland vegetation in Shetland'. *Scottish Geographical Magazine*, **106**: 28–36.

BIRSE, E. L. (1973). 'Vegetation of the Sullom Voe area, Shetland'. Macaulay Institute for Soil Research, Craigiebuckler, Aberdeen.

—— (1974). 'Bioclimatic characteristics of Shetland', *in* Goodier, R. (ed.), *The natural environment of Shetland*, 24–32. Nature Conservancy Council, Edinburgh.

BIRSE, E. L. & ROBERTSON, J. S. (1973). 'Report on the vegetation of the Baltasound area, Shetland'. Macaulay Institute for Soil Research, Craigiebuckler, Aberdeen.

BLACK, D. D. (1859). 'Notice of the excavation of a "Pict's House", on the farm of Stensall of Kergord, Zetland'. *Proceedings of the Society of Antiquaries of Scotland*, **2**: 452–454.

BLACKADDER, J. S. (1990). 'Rare berry is found in isles'. *Shetland Times*, 29 June 1990: 15. (A note on the hermaphrodite form of crowberry on Ronas Hill.)

—— (1991). 'Vanished member of isles' native flora will be restored—if cuttings take'. *Shetland Times*, 26 July 1991: 8. (Notes on a juniper which once grew near the Loch of Spiggie, South Mainland.)

—— (1992). 'Years of plant hunting pay off '. *Shetland Times*, 28 August 1992: 15. (Notes on juniper including the discovery of the intermediate form in Fetlar.)

—— (1993). [A note on Shetland junipers raised in Cambridge.] *Shetland Times*, 16 April 1993: 12.

—— (1994a). [A note of the discovery of juniper at Skelda Ness, West Mainland.] *Shetland Times*, 14 January 1994: 10.

—— (1994b). [A note on Shetland junipers raised in Cambridge.] *Shetland Times*, 8 July 1994: 15.

—— (1993). [A note on lesser water-parsnip.] *Shetland Times*, 7 May 1993: 12.

—— (1999a). [A note on the probable occurrence of French mallow in a garden in Lerwick.] *Shetland Times*, 24 September 1999: 20.

—— (1999b). [A note confirming the identity of French mallow in Lerwick.] *Shetland Times*, 15 October 1999: 15.

BLACKSTOCK, T. H. & ROBERTS, R. H. (1986). 'Observations on the morphology and fertility of *Juncus* × *surrejanus* Druce ex Stace & Lambinon in north-western Wales'. *Watsonia*, **16**: 55–63.

BLAEU, J. See Irvine (2006).

BÖCHER, T. W. (1943). 'Studies on Variation and Biology in Plantago lanceolata L.' *Dansk Botanisk Arkiv*, **11**: 1–18.

BOND, T. E. T. (1976). 'Length and insertion of the filaments in *Endymion*'. *Watsonia*, **11**: 141–142.

BÖRGESEN, F. (1908). 'Gardening and tree-planting', *in* Warming, E. (ed.), *Botany of the Færöes*, pt. 3: 1027–1043. Copenhagen.

BOSWELL, J. T. (1878). 'C[erastium] *triviale*, Link., d. *alpestre*'. *Botanical Exchange Club. Report for the year 1876*: 11.

BOTANICAL SOCIETY OF THE BRITISH ISLES (2007). *BSBI Recorder*, Spring 2007. (A fine photograph of the very rare pondweed hybrid, *Potamogeton* × *gessnacensis* at the Loch of Gards, Scatness, South Mainland, occupies the front cover.)

BRAITHWAITE, M. E., ELLIS, R. W. & PRESTON, C. D. (2006). *Change in the British Flora 1987–2004.* London.

BRAND, J. (1701). *A Brief Description of Orkney, Zetland, Pightland-Firth & Caithness.* Edinburgh. (An early reference to common scurvygrass on p. 80.)

BRATHAY EXPLORATION GROUP (1957). *Annual report and account of expeditions in 1956*: 30–31. (Brief notes on botanical studies in Foula in 1956.)

—— (1971). *Field Studies on Foula 1969 and 1970. Biology. Geography. Geology.* Field Studies Report, no. 11.

—— (1973). *Field Studies on Foula 1971 and 72.* Field Studies Report, no. 21.

—— (1974). *Field Studies on Foula 1973.* Field Studies Report, no. 27.

—— (1974). 'A note on the distribution of shore weed, *Littorella uniflora*. Foula August 1973'. *Brathay Exploration Group. Field Studies Report*, no. 27: 26–27.

BRIGGS, D. E. (1978). *Barley.* London.

BRIGGS, M. (1982). 'Falkland Islands daisies'. *B.S.B.I. News*, no. 32: 16–17.

BRITISH GIRLS' EXPLORING SOCIETY (*c.*1968). 'Expedition to the island of Fetlar, Shetlands—Summer 1967'. *British Girls' Exploring Society. Expedition Report 1967*: 7–52.

BRITTEN, J. (1869). '*Epilobium obscurum, Schreb.,* in Orkney or Shetland'. *Journal of Botany, British and foreign*, **7**: 340–341.

BRITTON, R. H. [1975]. 'A preliminary account of the macrophytic vegetation of the freshwaters of Shetland'. *Report to NCC on some aspects of the ecology of Shetland*, 119–129. Nature Conservancy Council, Edinburgh.

BROWN, N. E. (1892), comp. 'Cerastium, *Linn.*', *in* Boswell-Syme, J. T., *English Botany*, supplement to vols. **1**–**4** of ed. 3: 39–42. London.

BRUCE, G. (1871). 'On the improvement of natural pasture without tillage'. *Transactions of the Highland and Agricultural Society of Scotland*, **3** (fourth series) : 306–314.

BRUCE, [J.] (1908), publ. *Description of Ye Countrey of Zetland.* Privately published. (A collection of late seventeenth-century accounts of Shetland, many of which had already appeared in Sibbald (1711).)

BRUMMITT, R. K., KENT, D. H., LUSBY, P. S. & PALMER, R. C. (1987). 'The history and nomenclature of Thomas Edmondston's endemic Shetland *Cerastium*'. *Watsonia*, **16**: 291–297.

BRYDEN, J. (1841). 'United parishes of Sandsting and Aithsting'. *The Statistical Account of the Shetland Islands*, 97–144. Edinburgh.

BRYSTING, A. K. (2008). 'The arctic mouse-ear in Scotland—and why it is not arctic'. *Plant Ecology & Diversity*, **1**: 321–327.

BSBI FIELD MEETING, 1960. (1962), *in* Wallace, E. C. (comp.), 'Plant records'. *Proceedings of the Botanical Society of the British Isles*, **4**: 419–433.

BULLARD, E. R. (1985). 'Flowering plants and ferns', *in* Berry, R. J., *The Natural History of Orkney*, 250–261. London. (New Naturalist series, no. 70.)

———— (1995). *Wildflowers in Orkney. A new Checklist.* Kirkwall.

BULLARD, E. R. & GOODE, D. A. (1975). 'The vegetation of Orkney', *in* Goodier, R. (ed.), *The natural environment of Orkney*, 31–46. Nature Conservancy Council, Edinburgh.

BURDON, R. J. (1921), *in* Brown, G. C. (ed.), '*Report of the distributor for 1920*'. *Botanical Society and Exchange Club of the British Isles. Report for 1920*, **6**: 209–259.

———— (1922a), *in* Thomas, E. N. *et al.* (eds.), '*Report of the distibutors for 1920 [1921]*'. *Botanical Society and Exchange Club of the British Isles. Report for 1921*, **6**: 547–587.

———— (1922b), *in* Thompson, H. S. (ed.), *Thirty-eighth annual report of the Watson Botanical Exchange Club*, **3**: 178.

BURGESS, A. R. (1974). 'The distribution of aquatic plants in the Ham Burn system. Foula – August 1973'. *Brathay Exploration Group. Field Studies Report*, no. 27: 19–25.

BYERS, E. (1962), *in* Wallace, E. C. (comp.), 'Plant records'. *Proceedings of the Botanical Society of the British Isles*, **4**: 419–433.

CADBURY, C. J. (1961), *in* Wallace, E. C. (comp.), 'Plant records'. *Proceedings of the Botanical Society of the British Isles*, **4**: 162–172. (Cadbury records the discovery of sea pea at Nor Wick, Unst, on p. 165.)

CALDER, C. S. T. (1958). 'Report on the discovery of numerous stone age house-sites in Shetland'. *Proceedings of the Society of Antiquaries of Scotland*, **89**:

340–397. (The finding of carbonised barley grains on the site of a neolithic dwelling in West Mainland is discussed in some detail.)

CAMPBELL, J. (1774). '*Of the Islands of Shetland*', *in* Campbell, J., *A Political Survey of Britain*, **1**: 676–704. (Early references to juniper and rowan on p. 690.)

CARRICK, A., GRIGOR-TAYLOR, F., JOHNSON, A., MOYLAN, E. & SIGRUN, R. (1997). *Shetland Botanical Survey 1997.* [Lerwick.] (A report to the Shetland Islands Council.)

C[ARRUTHERS], W. [Notes on some plants collected in Shetland in 1865 by R. Tate]. *Journal of Botany, British and foreign*, **4**: 351. (See Tate (1866).)

CARTER, S. P., PROCTOR, J. & SLINGSBY, D. R. (1987a). 'Soil and vegetation of the Keen of Hamar serpentine, Shetland'. *Journal of Ecology*, **75**: 21–42.

——— (1987b). 'The Effects of Fertilization on Part of the Keen of Hamar Serpentine, Shetland'. *Transactions of the Botanical Society of Edinburgh*, **45**: 97–105.

CATTON, J. (1838). *The history and description of the Shetland Islands.* Wainfleet. (Primroses in the Cunningsburgh area, South Mainland, are noted on p. 71.)

CHAPELHOW, R. (1965). 'On glaciation in North Roe, Shetland'. *Geographical Journal*, **131**: 60–70.

CHARLTON, W. (2007), ed. *Travels in Shetland 1832-1852.* Lerwick. (An account of Edward Charlton's visits to Shetland in 1832, 1834, and 1852. In 1832 he mistakenly records maiden pink—not a Shetland plant—from near Lerwick.)

CHATER, A. O. & RICH, T. C. G. (1995). '*Rorippa islandica* (Oeder & Murray) Borbás (Brassicaceae) in Wales'. *Watsonia*, **20**: 229–238.

CHEFFINGS, C. M. & FARRELL, L. (2005), eds. *The Vascular Plant Red Data List for Great Britain.* Joint Nature Conservation Committee.

CHEYNE, J. R. (1949a). 'The Pattern of Da Voar'. *New Shetlander*, no. 15: 5–12.

——— (1949b). 'Roond Aboot Skeld in Time and Place', pt. 6. *New Shetlander*, no. 19: 27. (Cheyne mentions a number of local names for some common plants.)

CHITTENDEN, F. J. (1951), ed. *Dictionary of Gardening.* (Royal Horticultural Society.) Oxford.

CHRISTIANSEN, M. P. (1942). 'The Taraxacum-flora of Iceland'. *Botany of Iceland*, **3**, pt. 3: 229–343, plus forty-four plates.

CLAPHAM, A. R., TUTIN, T. G. & WARBURG, E. F. (1962). *Flora of the British Isles*, ed. 2. Cambridge.

CLEMENT, E. J. (1976). 'Adventive news 6'. *B.S.B.I. News*, no. 14: 11–16.

CLÉMENT, A. (1927). *J. G. Forchhammer. Rejse til Færøerne. Dagbog 28. April til 21. August 1821.* Copenhagen. (Forchhammer, the Holstein geologist who visited Scalloway on his way to Faeroe in 1821, noted the benefits of providing shelter for planted trees in the garden at Westshore, Scalloway (p. 17). This garden, which exists to this day, was mentioned earlier by Fothergill in his 1806 journal (p. 120), and by Hibbert (1822, p. 267–268).)

COLES, S. M. (1971). 'The *Ranunculus acris* L. complex in Europe'. *Watsonia*, **8**: 237–261.

COLVIN, S. (1899), ed. 'The letters of Robert Louis Stevenson'. *Scribner's Magazine*, **25**: 29–48. (Stevenson noted thrift 'among the lichened crevices of the old stone-work' of Fort Charlotte, Lerwick, in 1869 (p. 43).)

COMPTON, R. H. (1920). 'Melandryum dioicum', *in* Moss, C. E. (ed.), *Cambridge British Flora*, **3**: 72–73. Cambridge.

CONOLLY, A. P. (1977). 'The distribution and history in the British Isles of some alien species of *Polygonum* and *Reynoutria*'. *Watsonia*, **11**: 291–311.

COOK, C. D. K. (1961). '*Sparganium* in Britain'. *Watsonia*, **5**: 1–10.

—— (1975). '[*Ranunculus* L.]. Subg. *Batrachium* (DC.) A. Gray', *in* Stace, C. A. (ed.), *Hybridization and the Flora of the British Isles*, 125–128. London.

COPE, T. [A.] & GRAY, A. (2009). *Grasses of the British Isles*. B.S.B.I. Handbook, no. 13.

COPE, T. A. & STACE, C. A. (1978). 'The *Juncus bufonius* L. aggregate in western Europe'. *Watsonia*, **12**: 113–128, plus two plates.

COPLAND, J. (1960), *in* Wallace, E. C. (comp.), 'Plant records'. *Proceedings of the Botanical Society of the British Isles*, **4**: 46–54.

—— (1996). *Hardy plants in the North*, ed. 3. Brae, Shetland. (In the 1980s, Copland, 'a skilled gardener and plantsman' (Steppanova (2004)), established a successful nursery at Islesburgh, Northmavine.)

COUTTS, J. (2003). 'Fetlar Biodiversity Action Plan'. >www.livingshetland.org.uk/ documents/FetlarCommunityBAP.pdf< Last accessed 18 February 2011.

COWIE, R. (1871). *Shetland: descriptive and historical*, ed. 1. Aberdeen. (A reference to sycamores at Sound, Weisdale, Central Mainland, on p. 282.)

—— (1874). *Shetland: descriptive and historical*, ed. 2. Edinburgh. (A reference to wild roses at Mavis Grind, Northmavine, on p. 243.)

CRABBE, J. A., JERMY, A. C. & WALKER, S. (1970). 'The Distribution of *Dryopteris assinilis* S. Walker in Britain'. *Watsonia*, **8**: 3–13.

CRAIG-CHRISTIE, A. (1870). 'Notes of a Botanical Excursion to Shetland in 1868'. *Transactions and Proceedings of the Botanical Society* [of Edinburgh], **10**:165–170.

CRAWSHAW, B. (1971). 'The distribution of Eleocharis and Littorella in Mill Loch. August 1970'. *Brathay Exploration Group. Field Studies Report*, no. 11: 77–78. (A brief account of these two genera in the Mill Loch, Foula.)

CURRIE, A. (1960). 'Further notes on the flora of Fair Isle (V. c. 112)'. *Proceedings of the Botanical Society of the British Isles*, **4**: 38–39.

CURSITER, J. W. (1894). *List of books and pamphlets relating to Orkney and Shetland*. Kirkwall.

DAHLSTEDT, H. (1903). 'The Hieracia from the Færöes', *in* Warming, E. (ed.), *Botany of the Færöes*, pt. 2: 625–659, plus two plates. Copenhagen.

—— (1930). 'De Svenska arterna av släktet Taraxacum. VIII. Spectabilia'. *Kungliga Svenska Vetenskapsakademiens Handlingar*, **9** (third series): 1–97.

DALBY, D. H. (1976). 'The salt marshes of Sullom Voe'. (A report to the Sullom Voe Association.)

—— (1981). 'The salt marshes of Sullom Voe'. *Proceedings of the Royal Society of Edinburgh*, **80** (Section B. Biological Sciences): 191–202.

—— (1985). 'Salt-marsh vegetation in the Shetland Islands'. *Vegetatio*, **61**: 45–54.

—— (1986). 'Shetland salt marsh survey, 1985'. (A report to the Nature Conservancy Council.)

—— (1987a). '*Atriplex littoralis* new to Shetland'. *Watsonia*, **16**: 330–331.

—— (1987b). 'Identification of Vegetative *Triglochin* Plants'. *Transactions of the Botanical Society of Edinburgh*, **45**: 127–129.

—— (1995). 'The use of *Puccinellia maritima* in salt marsh monitoring'. *Proceedings of the Royal Society of Edinburgh*, **103** (Section B. Biological Sciences): 219–231.

DALBY, D. H. & D ALBY, C. (1989). 'An Overlooked Taxonomic Character in *Eleocharis multicaulis'*. *Transactions of the Botanical Society of Edinburgh*, **45**: 319–322.

DALRYMPLE, S. & CHAMBERS, C. (2010). 'Morphological variation and spatial separation of two races of *Cerastium nigrescens* (Shetland Mouse-ear)'. *BSBI News*, no. 113: 45.

DANDY, J. E. & TAYLOR, G. (1938). 'Studies of British Potamogetons. — III. *Potamogeton rutilus* in Britain'. *Journal of Botany, British and foreign*, **76**: 239–241.

—— (1940a). 'Studies of British Potamogetons. — XII. *Potamogeton pusillus* in Great Britain'. *Journal of Botany, British and foreign*, **78**: 1–11.

—— (1940b). 'Studies of British Potamogetons. — XIII. *Potamogeton Berchtoldii* in Great Britain'. *Journal of Botany, British and foreign*, **78**; 49–66.

—— (1940c). 'Studies of British Potamogetons. — XIV. *Potamogeton* in the Hebrides (Vice-county 110)'. *Journal of Botany, British and foreign*, **78**: 139–147.

DAVIDSON, W. D. (1938). *Potato growing for seed purposes*. Dublin.

DAVIES, E. W. (1953). 'Notes on *Carex flava* and its allies. I – A sedge new to the British Isles'. *Watsonia*, **3**: 66–69.

DAVIES, E. W. & PADMORE, P. A. (1954), *in* Wallace, E. C. (comp.), 'Plant records'. *Proceedings of the Botanical Society of the British Isles*, **1**: 161–185.

DE BOVET, M. A. (1898). *L'Ecosse*. Paris. (Contains, on p. 244, a list of plants from which wool dyes are said to have been obtained.)

D[ENNES], G. E. (1845). [A note on *Cerastium nigrescens*]. *The Phytologist: a popular botanical miscellany*, **2**: 95–96.

DENNIS, M. T. (1966). 'Notes on the Flora, 1966'. *Fair Isle Bird Observatory. Bulletin*, **5** (new series): 225–226.

DRABBLE, E. (1931). 'The British forms of *Ranunculus acer* L.'. *Botanical Society and Exchange Club of the British Isles. Report for 1930*, **9**: 472–478.

DRABBLE, H. (1937), comp. 'Pansy records'. *Botanical Society and Exchange Club of the British Isles. Report for 1936*, **11**: 319–324.

DRUCE, G. C. (1911). 'The alpine Cerastia of Britain'. *Annals of Scottish Natural History*, [**20**]: 38–44.

—— (1915a), *in* Druce, G. C. (ed.), 'Plant notes, etc., for 1914'. *Report of the Botanical Society and Exchange Club of the British Isles for 1914*, **4**: 7–48. (The first published British record, on p. 17, of the hybrid *Senecio aquaticus* × *S. jacobaea*, found at Scalloway in 1888 by W. H. Beeby.)

—— (1915b), *in* Druce, G. C. (ed.), 'New county and other records'. *Report of the Botanical Society and Exchange Club of the British Isles for 1914*, **4**: 59–79.

—— (1921a), *in* Druce, G. C. (ed.), 'Plant notes, etc., for 1920'. *Botanical Society and Exchange Club of the British Isles. Report for 1920*, **6**: 14–57.

—— (1921b), *in* Druce, G. C. (ed.), 'New county and other records'. *Botanical Society and Exchange Club of the British Isles. Report for 1920*, **6**: 108–163.

—— (1921c), *in* Brown, G. C. (ed.), '*Report of the distributor for 1920*'. *Botanical Society and Exchange Club of the British Isles. Report for 1920*, **6**: 209–259.

—— (1921d). 'Shetland Plants'. *Proceedings of the Linnean Society of London*, [**133**]: 77–78.

—— (1922a), *in* Druce, G. C. (ed.), 'Plant notes, etc., for 1921'. *Botanical Society and Exchange Club of the British Isles. Report for1921*, **6**: 272–327.

—— (1922b), *in* Druce, G. C. (ed.), 'New county and other records'. *Botanical Society and Exchange Club of the British Isles. Report for 1921*, **6**: 369–404.

—— (1922c). 'Flora Zetlandica'. *Botanical Society and Exchange Club of the British Isles. Report for 1921*, **6** (supplement to pt. 3): 457–546. Also issued as a separately paginated offprint; Druce's own annotated copy, in the Fielding-Druce Herbarium, Department of Plant Sciences, Oxford, is one of the latter.

—— (1922d), *in* Thomas, E. N. *et al.* (eds.), '*Report of the distributors for 1920* [*1921*]'. *Botanical Society and Exchange Club of the British Isles. Report for 1921*, **6**: 547–587.

—— (1923), *in* Druce, G. C. (ed.), 'Plant notes, etc., for 1922'. *Botanical Society and Exchange Club of the British Isles. Report for 1922*, **6**: 600–640.

—— (1925a), *in* Druce, G. C. (ed.), 'Plant notes, etc., for 1924'. *Botanical Society and Exchange Club of the British Isles. Report for 1924*, **7**: 431–458.

—— (1925b), *in* Druce, G. C. (ed.), 'New county and other records'. *Botanical Society and Exchange Club of the British Isles. Report for 1924*, **7**: 552–605.

—— (1925c). 'Additions to the Flora Zetlandica'. *Botanical Society and Exchange Club of the British Isles. Report for 1924*, **7**: 628–657.

—— (1926a), *in* Druce, G. C. (ed.), 'Plant notes, etc., for 1925'. *Botanical Society and Exchange Club of the British Isles. Report for 1925*, **7**: 762–793.

—— (1926b), *in* Druce, G. C. (ed.), 'New county and other records'. *Botanical Society and Exchange Club of the British Isles. Report for 1925*, **7**: 859–908.

—— (1926c), *in* Druce, G. C. (ed.), 'Plants new to Britain'. *Botanical Society and Exchange Club of the British Isles. Report for 1925*, **7**; 996–998. (On p. 997 there is a reference to the hawkweed *Hieracium australius* under the heading '*H. polycomum*'.)

—— (1929), *in* Druce, G. C. (ed.), 'Plant notes, etc., for 1928'. *Botanical Society and Exchange Club of the British Isles. Report for 1928*, **8**: 608–641.

—— (1930a), *in* Druce, G. C. (ed.), 'Plant notes, etc., for 1929'. *Botanical Society and Exchange Club of the British Isles. Report for 1929*, **9**: 18–44.

—— (1930b), *in* Druce, G. C. (ed.), 'New county and other records'. *Botanical Society and Exchange Club of the British Isles. Report for 1929*, **9**: 100–147.

—— (1932). *The Comital Flora of the British Isles*. Arbroath.

DUDMAN, A. A. & RICHARDS, A. J. (1994). 'Seven new species of *Taraxacum* Wigg. (Asteraceae), native to the British Isles'. *Watsonia*, **20**: 119–132.

—— (1997). *Dandelions of Great Britain and Ireland*. B.S.B.I. Handbook, no. 9.

EDMONDSTON, A. (1809). *A view of the ancient and present state of the Zetland Islands*, **1**. Edinburgh. (Contains comments on crops and cultivation (p. 195–211), and on trees (p. 211–219).)

[EDMONDSTON, E.] (1868). *The young Shetlander: Shadow over the Sunshine.* Edinburgh. (A biography of Thomas Edmondston, the youthful Unst botanist, edited by his mother.)

EDMONDSTON, L. (1841). 'General observations on the county of Shetland', in *The Statistical Account of the Shetland Islands*, 145–175. Edinburgh.

EDMONDSTON, T. Commenced November 1837. 'Flora Shetlandica No 1'. Incomplete MS in the library of the Natural History Museum, Cromwell Road, South Kensington, London.

—— (1839). 'List of Plants observed in the Island of Unst, Shetland, during the summer of 1837', *in* Hooker, W. D., *Notes on Norway*, ed. 2: 111–117. Privately published. (An important work which contains the first published list of Shetland plants.)

—— (1840). '*The Ligneous Flora of the Shetland Islands*'. *Gardener's Magazine*, **16**: 102.

—— (1841a). '*List of Phanerogamous Plants, together with the Cryptogamic Orders* Filices, Equisetaceae, *and* Lycopodiaceae, *observed in the Shetland Islands*'. *Annals and Magazine of Natural History*, **7**: 287–295.

—— (1841b). 'On the native dyes of the Shetland Islands'. *Transactions of the Botanical Society* [of Edinburgh], **1**: 123–126.

—— (1841c). 'Catalogue of the Phaenogamous plants and ferns observed in the Shetland Islands', *in* Edmondston, L., 'General observations on the county of Shetland', in *The Statistical Account of the Shetland Islands*, 145–175. Edinburgh. (T. Edmondston's catalogue is on p. 150–153.)

—— (1842a). 'Remarks on the Flora of Shetland, with a full catalogue of plants observed in these islands'. *Annals and Magazine of Natural History*, **9**; 69–70.

—— (1842b). '. . . on the The Botany of Shetland'. *The Phytologist: a popular botanical miscellany*, **1**: 430–432. (Repeated, almost verbatim, in *Annals and Magazine of Natural History*, **11** (1843): 70–73.)

—— (1843a). '*Notice of a new British Cerastium*'. *The Phytologist: a popular botanical miscellany*, **1**: 497–500.

—— (1843b). '*Note on Cerastium latifolium*'. *The Phytologist: a popular botanical miscellany*, **1**: 677–678.

—— (1843c). '*Shetland locality for Cynosurus echinatus*'. *The Phytologist: a popular botanical miscellany*, **1**: 772.

—— (1844). 'Remarks on the Botany of Shetland'. *Transactions of the Botanical Society* [of Edinburgh], **1**: 185–188. (Based on T. Edmondston (1842b).)

—— (1845a). *A Flora of Shetland.* Aberdeen.

—— (1845b). '*A few Parting Notes*'. *The Phytologist: a popular botanical miscellany*, **2**: 182–184. (In these notes Edmondston mentions his being 'most unexpectedly and suddenly summoned to go to the west coast of America', as naturalist on HMS 'Herald', a journey which ended in his accidental death in 1846 in South America. See Seemann (1853).)

EDWARDS, A. M. (1977). 'Alpines and other flowers of the Shetland Isles'. *Quarterly Bulletin of the Alpine Garden Society*, **45**: 297–308.

ELDER, F. & RIBBONS, B. W. (1962), *in* Wallace, E. C. (comp.), 'Plant records'. *Proceedings of the Botanical Society of the British Isles*, **4**: 419–433.

ERDTMAN, G. (1924). 'Studies in the Micropalæontology of Postglacial Deposits in Northern Scotland and the Scotch Isles, with especial reference to the history of the woodlands'. *Journal of the Linnean Society. Botany*, **46**: 449–504, plus one plate.

EVANS, A. H. & BUCKLEY, T. E. (1899). *A vertebrate fauna of the Shetland Islands*. Edinburgh.

EVERSHED, H. (1874). 'On the agriculture of the islands of Shetland'. *Transactions of the Highland and Agricultural Society of Scotland*, **6** (fourth series) : 187–228.

—— (1884). '*Woods and wilds of Shetland* '. *Forestry*, **8**: 252–256. (Evershed's mention of 'Cranberry' on p. 254 clearly refers to crowberry.)

EXPERIMENTAL CARTOGRAPHY UNIT (1978). 'Maps from the Shetlands data bank'. *Geographical Magazine*, **50**: 736–753.

FARRELL, L. (1977). 'Brathay expedition to Ronas Hill, Shetland, or a botanist's summer holiday'. *B.S.B.I. News*, no. 17: 12–13.

—— (1978). 'Report on the Kergord Plantations. August 1977'. (A report to the Nature Conservancy Council.)

—— (1996). 'Shetland (v. c. 112). 1st–7th August'. *BSBI News*, no. 71: 70–72. (An account of the BSBI field meeting, 1995.)

FEILBERG, P. (1900). *Fra Lier og Fjelde, breve til hjemmet 1899*. Copenhagen. (Contains a number of references to common Shetland species.)

—— (1908). 'Some notes on the agriculture of the Færöes', *in* Warming, E. (ed.), *Botany of the Færöes*, pt. 3: 1044–1054. Copenhagen.

FENTON, A. (1978). *The northern isles: Orkney and Shetland*. Edinburgh.

FERREIRA, R. E. C. (1963). 'Some Distinctions Between Calciphilous and Basiphilous Plants. I. Field Data'. *Transactions and Proceedings of the Botanical Society of Edinburgh*, **39**: 399–413.

FIRTH, H. (1994). 'Angelica and Norsemen'. *Orkney Naturalist*, 1–7. (Bulletin of the Orkney Field Club, 1994.)

FITTER, R. S. R. (1959). 'Addenda to the flora of Fair Isle'. *Proceedings of the Botanical Society of the British Isles*, **3**: 172–173.

FLINN, D. (1974). 'The coastline of Shetland', *in* Goodier, R. (ed.), *The natural environment of Shetland*, 13–23. Nature Conservancy Council, Edinburgh.

[FLINT, P.] (2008). 'Unusual little find'. *Shetland Times*, 22 August 2008: 37.

FORCHHAMMER, J. G. See Clément (1927).

FOTHERGILL, C. (1806a). 'Voyage to the Orcades & Shetland', **3**. MS in Shetland Museum and Archives, Hay's Dock, Lerwick, Shetland. (Contains a reference to English stonecrop on Bruray, in the Out Skerries, on p. 102.)

—— (1806b). 'A View of the Natural History of the Isles In a Series of Descriptive Catalogues of the Quadrupeds Birds Fish Insects Reptiles Shells Crustaceous Animals Zoophytes and Plants That have hitherto been discovered in the Orcades, Shetland, Fair Isle, and Fula'. MS in the Thomas Fisher Rare Book Library, University of Toronto, Canada. Copy in Shetland Museum and Archives, Hay's Dock, Lerwick, Shetland.

FOWLER, J. A. (1978), ed. 'Preliminary studies on the fauna and flora of Yell, Shetland'. (Report of the Leicester Polytechnic (School of Life Sciences) expedition to Shetland, July 1977.)

FRASER, D. (1911). *Angling sketches from a wayside inn.* Edinburgh. (On p. 153 there is a reference to 'wild harebell' in the Loch of Watlee area , Unst.)

[FRASER, J.] (1934), *in* Pearsall, W. H. (ed.), 'Plant notes for 1933'. *Botanical Society and Exchange Club of the British Isles. Report for 1933*, **10**: 468–487. (Contains a description of false apple-mint on p. 479–480.)

FRASER-JENKINS, C. R. (2007). 'The species and subspecies in the Dryopteris affinis group'. *Fern Gazette*, **18**: 1–26.

GAMMACK, J. J. M. (1982). 'July's Nature Walks'. *Shetland Life*, no. 22: 37–39. (A note about marram on the Isle of Noss.)

GAMMACK, J. J. M. & RICHARDSON, M. G. (1980). *A Compendium of Ecological and Physical Information on the Shetland Coastline.* Nature Conservancy Council.

GARDIE HOUSE PAPERS. 1674. 'Information anent Magnus Laurenson in Gonfirth, sorcerer'. Laurenson claimed to be able to cure many illnesses, and used, in some cases, plants from the wild. One such plant mentioned in this document, 'limbrick grasse', is the first known reference to bog asphodel in Shetland. The identity of the few others listed is open to doubt. Document in 'Gardie House', Bressay, Shetland.

GARDINER, J. C. (1960), *in* Wallace, E. C. (ed.), 'Plant records'. *Proceedings of the Botanical Society of the British Isles*, **3**: 394–419. (An erroneous reference to Borrer's saltmarsh-grass on p. 417.)

GEAR, S. (1981). 'Foula Diary'. *Shetland Life*, no. 12: 22–23.

——— (2008). *Flora of Foula.* Foula, Shetland. Foula Heritage. (The first comprehensive pictorial guide to the wild plants of one of our smaller islands.)

GILL, J. J. B. (1971). 'Cochlearia scotica Druce—Does it exist in northern Scotland?'. *Watsonia*, **8**: 395–396.

GILL, J. J. B., McALLISTER, H. A. & FEARN, G. M. (1978). 'Cytotaxonomic studies on the *Cochlearia officinalis* L. group from inland stations in Britain'. *Watsonia*, **12**: 15–21.

GIMINGHAM, C. H., WELCH, D., CLEMENT, E. J. & LANE, P. (2002). 'A large population of *Plagiobothrys scouleri* (Boraginaceae) in north-east Scotland, and notes on occurrences elsewhere in Britain'. *Watsonia*, **24**: 159–169.

GLADSTONE, J. See Turrill (1929).

GODDARD, P. (2003). 'Island of Bressay, Shetland. Community Biodiversity Action Plan'. >http://www.livingshetland.org.uk/documents/ BressayCommunityBAP.pdf< Last accessed 18 February 2011.

GODFREY, R. (1897). 'Unst and its bird life'. *The Field*, 30 January 1897: 134–135. (An erroneous reference to 'rock rose' on p. 134.)

GOLDSMITH, F. B.(1974). 'The sea-cliff vegetation of Shetland'. Discussion Papers in Conservation, no. 9. University College, London. (Also, slightly altered, in *Journal of Biogeography*, **2** (1975): 297–308.)

GOODE, D. A. (1974). 'The flora and vegetation of Shetland', *in* Goodier, R. (ed.), *The natural environment of Shetland*, 50–72. Nature Conservancy Council, Edinburgh.

GOODE, D. A. & FIELD, E. M. (1973). 'Notes on the vegetation of Shetland bogs'. Nature Conservancy Council.

GOODIER, R. (1974), ed. *The natural environment of Shetland.* Nature Conservancy Council, Edinburgh.

GORDON, J. (1794). 'United Parishes of North Yell and Fetlar', *in* Sinclair, J. (comp.), *The statistical account of Scotland*, **13**: 278–291. (Hazel, rowan, and elder [perhaps intending alder] are reported from the N of Yell and/or Fetlar.)

GORNALL, R. J. (1988). 'The coastal ecodeme of *Parnassia palustris* L.'. *Watsonia*, **17**: 139–143.

GORNALL, R. J. & WENTWORTH, J. E. (1993). 'Variation in the chromosome number of *Parnassia palustris* L. in the British Isles'. *New Phytologist*, **123**: 383–388.

G[ORRIE], W. (1855). '*D[actylis] caespitosa*', *in* Morton, J. C. (ed.), *A Cyclopedia of Agriculture, practical and scientific*, **1**: 601–602. Glasgow. (Contains much interesting information on tussac-grass, and on the early attempts to introduce it into Britain.)

GRAHAM, [R.] (1838). 'Observations by Dr. Graham on Plants collected in Scotland in 1837 by Dr. M'Nab ...'. *Annals of Natural* History, **1**: 406–408.

GRAHAM, G. G. & PRIMAVESI, A. L. (1993). *Roses of Great Britain and Ireland.* B.S.B.I. Handbook, no. 7.

GRAVES, F. S. (1899). 'Wilson's Filmy Fern on Foula'. *Annals of Scottish Natural History*, [**8**]: 243.

GREIG, P. W. (1892). *Annals of a Shetland parish: Delting.* Lerwick. (On p. 69 there are references to rowan and wild roses at the Burn of Valayre, near Brae.)

[GROAT, A. G.] (1831). *Thoughts on Orkney and Zetland.* Edinburgh. (Includes lists of books and MMS relevant to Orkney and/or Shetland.)

GUSTAFSSON, M. (1976). 'Evolutionary trends in the Atriplex prostrata group of Scandinavia. Taxonomy and morphological variation'. *Opera Botanica*, no. 39.

HAGERUP, O. (1927). '*Empetrum hermaphroditum* (Lge) Hagerup. A new tetraploid, bisexual species'. *Dansk Botanisk Arkiv*, **5**: 1–17.

——— (1933). 'Studies on polyploid ecotypes in Vaccinium uliginosum'. *Hereditas*, **18**: 122–128.

HALDANE, D. & HARDIE, H. G. M. (1943). *Limestones of Scotland. Area VIII. Orkney and Shetland.* Department of Scientific and Industrial Research (Geological Survey of Great Britain). Wartime pamphlet no. 13.

HAMBLER, D. J. (1958). 'Some taxonomic investigations on the genus *Rhinanthus*'. *Watsonia*, **4**: 101–116.

HANBURY, F. J. (1888). 'Notes on some Hieracia new to Britain'. *Journal of Botany, British and foreign*, **26**: 204–206.

——— (1894). 'Notes on British Hieracia'. *Journal of Botany, British and Foreign*, **32**: 225–233.

——— (1895). 'Notes on the *Lepidoptera* observed during a short botanical tour in West Sutherland, the Orkneys, and Shetlands'. *Entomologist's Monthly Magazine*, **6** (second series): 1–12.

HANSEN, K. (1966). 'Vascular Plants in the Faeroes'. *Dansk Botanisk Arkiv*, **24**: no. 3.

HARDING, M. (1987). 'The interaction of salt and grazing on cliff top pastures in Shetland'. (Part of an M.Sc. ecology degree at the University of Aberdeen.)

HARDY, E. [1950]. 'Our Flowers', in *As far as you can* go ... . (Lerwick and Shetland. The Official Guide.)

HARGREAVES, S. (2009). 'Assessment of *Trifolium* spp. across the UK—the importance of outer islands'. *BSBI News*, no. 111: 42–43.

HARVEY, P., LOCKTON, A. & WHILD, S. (2007). 'Shetland (v. c. 112), 17th–20th July'. *BSBI News*, no. 105: 39–40. (An account of the BSBI field meeting, 2006.)

HARROLD, P. (1978). 'A Glabrous Variety of Sagina subulata (Sw.) C. Presl in Britain'. *Transactions of the Botanical Society of Edinburgh*, **43**: 1–5.

HARVIE-BROWN, J. A. (1892). 'Hedgehog (*Erinaceus europæus*) in Shetland'. *Annals of Scottish Natural History*, [**1**]: 132. (A note on the introduction of gorse in the Tingwall area, Central Mainland.)

HAUKE, R. L. (1967). 'A Systematic Study of Equisetum arvense'. *Nova Hedwigia*, **13**: 81–109, plus nine plates.

HAWKSWORTH, D. L. (1966a). 'Foula—botanical studies & preliminary report on the peat deposits'. *Brathay Exploration Group. Annual report and account of expeditions in 1965*: 38–42.

——— (1966b). 'Foula—literature on the botany, geography & geology'. *Brathay Exploration Group. Annual report and account of expeditions in 1965*, 43–45.

——— (1967). 'Foula—botanical studies: 2'. *Brathay Exploration Group. Annual report and account of expeditions in 1966*: 139–146.

——— (1969). 'Notes on the flora and vegetation of Foula, Zetland (V. C. 112)'. *Proceedings of the Botanical Society of the British Isles*, **7**: 537–547.

——— (1970). 'Studies on the Peat Deposits of the Island of Foula, Shetland'. *Transactions and Proceedings of the Botanical Society of Edinburgh*, **40**: 576–591.

——— (1971). 'Two additions to the flora of Foula, Zetland, v. c. 112'. *Watsonia*, **8**: 301.

HENDERSON, D. M. & DICKSON, J. H. (1994). *A naturalist in the Highlands.* Edinburgh. (An account of James Robertson whose travels in Shetland in 1769 provides us with the earliest listing of the plants of our area.)

HEPPER, F. N. (1975). '*Sagina* L.', in Stace, C. A. (ed.), *Hybridization and the Flora of the British Isles*, 175–176. London.

HESLOP-HARRISON, Y. (1953). 'Variation in *Nymphaea alba* in the British Isles'. *Botanical Society of the British Isles. Year Book, 1953*, 58–59.

HEWITT, J. (2009). 'Making an herbarium accessible'. *SLBI Gazette*, no. 8 (second series): 6–7. (An article on the Palmer-Scott collection at the South London Botanical Institute, 323 Norwood Road, London. Beeby's extensive Shetland herbarium is in the same institution.)

HIBBERT, S. (1822). *A Description of the Shetland Islands.* Edinburgh. (A reference to 'marigold' (perhaps intending marsh-marigold) appears on p. 442.)

HILLIAM, J. (1977). 'Phytosociological Studies in the Southern Isles of Shetland'. (A thesis submitted for a Ph.D. degree in the University of Durham.)

HINTON, M. P. (1991). 'Weed associates of recently grown *Avena strigosa* Schreber from Shetland, Scotland'. *Circaea*, **8**: 49–54.

HØEG, O. A. (1941). 'Jonsokgras, Plantago lanceolata'. *Det Kongelige Norske Videnskabers Selskabs. Forhandlinger*, **13**: 157–160.

HOLBOURN, S. C. (1961). 'Further Additions to the Flora of Foula'. *Transactions and Proceedings of the Botanical Society of Edinburgh*, **39**: 235–236.

HOLMES, N. T. H. (1979). '*A guide to identification of Batrachium* Ranunculus *species of Britain*'. *CST Notes*, no. 14. Nature Conservancy Council, London.

HOOKER, J. D. (1870). *The Student's Flora of the British Islands*. London. (On p. 272 a reference to '*Euphrasia officinalis* var. *maritima*' from the Shetland coast may be an early allusion to *E. foulaensis*.)

HOOKER, W. D. See Edmondston, T. (1839).

HOOKER, W. J. (1821). *Flora Scotica*. London. (On p. 17 of pt. 1 there is a record of saltmarsh flat-sedge from Bressay.)

—— (1838). *The British flora*, ed. 4, **1**. London. (A reference to the discovery of arctic sandwort in Unst appears on p. 182.)

HOPPE, G. (1965). 'Submarine peat in the Shetland Islands'. *Geografiska Annaler*, **47** (series A): 195–203.

HOWIE, A. (1945). 'Agriculture in Shetland'. *Scottish journal of agriculture*, **25**: 87–94. (An interesting account of the agricultural scene in Shetland around the time of the Second World War.)

HOWISON, [W.] (1829). '*Gardening in the Shetland Islands*'. *Gardener's Magazine*, **5**: 663–664.

HUNTER, F. (1882). [A note about common bistort near Lerwick.] *Botanical Record Club. Report for the year 1880*: 137.

INSTITUTE OF TERRESTRIAL ECOLOGY (1975). 'Report to NCC on some aspects of the ecology of Shetland'. (A report to the Nature Conservancy Council.)

IRVINE, J. M. (2006), ed. *The Orkneys and Schetland in Blaeu's Atlas Novus of 1654*. Published by the author.

IRVINE, T. 1853. 'Journal of Thomas Irvine 1853'. MS in Shetland Museum and Archives, Hay's Dock, Lerwick, Shetland. (Contains references to native shrubs.)

JACKSON, G. (1971a). 'The distribution of Potamogeton and Myriophyllum in Mill Loch. August 1970'. *Brathay Exploration Group. Field Studies Report*, no. 11: 74–76. (Concerns Mill Loch, Foula.)

—— (1971b). 'The distribution of Potamogeton, Sparganium, Myriophyllum in lochs, pools and streams on Foula. August, 1970'. *Brathay Exploration Group. Field Studies Report*, no. 11: 79–81.

—— (1974). 'Notes on the distribution and habitat of *Hymenophyllum wilsonii* (Wilson[']s Filmy Fern) on Foula'. *Brathay Exploration Group. Field Studies Report*, no. 27: 28–31.

JAKOBSEN, J. (1901). 'Shetlandsøernes stednavne'. *Aarbøger for Nordisk oldkyndighed og historie*, **16**. (Jakobsen lists a number of Norn place-names, on p. 137, indicative of the presence of wild roses.)

JAY, S. C. [1995]. 'Shetland[']s Native Trees'. Shetland Amenity Trust, Lerwick.

JEAVONS, R. [1970]. 'Botany'. *The 1969 report of the Schools Hebridean Society*, 82–89.

JERMY, A. C., ARNOLD, H. R., FARRELL, L. & PERRING, F. H. (1978). *Atlas of ferns of the British Isles*. London.

JERMY, [A.] C., SIMPSON, D., FOLEY, M. & PORTER, M. (2007). *Sedges of the British Isles*. B.S.B.I. Handbook, no. 1 (ed. 3).

JÓHANSEN, J. (1975). 'Pollen diagrams from the Shetland and Faroe Islands'. *New Phytologist*, **75**: 369–387.

—— (1978). 'The age of the introduction of *Plantago lanceolata* to the Shetland Islands'. *Danmarks Geologiske Undersøgelse. Årbog 1976*, 45–48.

—— (1985a). 'Man's first arrival in Shetland'. *Shetland Life*, no. 51: 4–6.

—— (1985b). 'Studies in the vegetational history of the Faroe and Shetland Islands'. *Annales Societatis Scientiarum Færoensis*, supplement 11.

JOHNSTON, H. H. 1926–1929. [Field notebooks covering his visits to Fetlar in these years.] Orkney Museum, Broad Street, Kirkwall, Orkney.

—— (1927). 'Additions to the Flora of Shetland'. *Transactions and Proceedings of the Botanical Society of Edinburgh*, **29**: 429–430. Also issued separately as the first paper in a series of four very short papers bearing the same title.

—— (1928). 'Additions to the Flora of Shetland', second paper. Privately published.

—— (1929a). 'Additions to the Flora of Shetland', third paper. Privately published.

—— (1929b). 'Additions to the Flora of Shetland', fourth paper. Privately published.

—— (1930), *in* Druce, G. C. (ed.), 'New county and other records'. *Botanical Society and Exchange Club of the British Isles. Report for 1929*, **9**: 100–147.

JOHNSTON, J. L. (1974). 'Shetland habitats, an outline ecological framework', *in* Goodier, R. (ed.), *The natural environment of Shetland*, 33–49. Nature Conservancy Council, Edinburgh.

—— (2007). *Victorians 60° North*. Lerwick. (The story of the Edmondstons and Saxbys of Shetland.)

KAY, Q. O. N. (1969). 'The origin and distribution of diploid and tetraploid *Tripleurospermum inodorum* (L.) Schultz Bip'. *Watsonia*, **7**: 130–141.

—— (1972). 'Variation in sea mayweed (*Tripleurospermun maritimum* (L.) Koch in the British Isles'. *Watsonia*, **9**: 81–107.

KAY, S. & PROCTOR, J. (2003). 'Population Dynamics of Two Scottish Ultramafic (Serpentine) Rarities with Contrasting Life Histories'. *Botanical Journal of Scotland*, **55**: 269–285.

KENT, D. H. (1992). *List of vascular plants of the British Isles*. London.

KENT, D. H. & ALLEN, D. E. (1984), comps. *British and Irish herbaria*. London.

KILLICK, J. (2006). 'Richard Charles Palmer 1935–2005'. *Watsonia*, **26**: 102–103. (An obituary.)

KIRKPATRICK, A. H. & MACDONALD, A. J. (1997). '*Calluna vulgaris* (L.) Hull dieback in Orkney, Scotland'. *Watsonia*, **21**: 271–275.

LAING, J. (1815). *An account of a voyage to Spitzbergen*. London. (A reference to juniper in Shetland on p. 26.)

LANSDOWN, R. V.(2008). *Water-starworts* (Callitriche) *of Europe*. B.S.B.I. Handbook, no. 11.

LATTER, D. & LEITCH, S. (1967). 'Botany'. *British Girls' Exploring Society. Foula Expedition 1966*, 6–16.

LAWSON, P. & SON (1836). 'Plants cultivated chiefly for their roots. 1. Plants having tuberous roots'. *Agriculturalist's manual*, 213–232. London. (Provides short descriptions of numerous potato varieties.)

LERWICK SHERIFF COURT RECORDS. 1771–1772. (SC12/6/1771/3). [A
    reference to the cutting of 'floss' (soft-rush) on the Clift Hills, South Mainland, in
    1771, resulting in an altercation between people from Cunningsburgh and
    West Burra.] Shetland Museum and Archives, Hay's Dock, Lerwick, Shetland.

LEWIS, A. M. (1976). 'Phytosociological studies in the northern isles of Shetland'.
    (A thesis submitted for a Ph.D. degree in the University of Durham.)

LEWIS, F. J. (1907). 'The Plant Remains in the Scottish Peat Mosses. Part III. The
    Scottish Highlands and the Shetland Islands'. *Transactions of the Royal
    Society of Edinburgh*, **46**: 33–70, plus four plates.

—— (1911). 'The Plant Remains in the Scottish Peat Mosses. Part IV. The
    Scottish Highlands and Shetland, with an Appendix on the Icelandic Peat
    Deposits'. *Transactions of the Royal Society of Edinburgh*, **47**: 793–833, plus
    five plates.

LIGHTFOOT, J. (1777). *Flora Scotica*, **2**. London. (On p. 544–545, plus one plate,
    there is an account of curved sedge, including a note of its occurrence in Shetland.)

LINDBERG, H. (1909). 'Die nordischen Alchemilla vulgaris-Formen und ihre
    Verbreitung'. *Acta Societatis Scientiarum Fennicæ*, **37**. (Brief references on p. 98
    to five stations for lady's-mantle by W. H. Beeby.)

LINTON, W. R. (1886). 'New records'. *Journal of Botany, British and foreign*, **24**:
    376–377.

—— (1905). *An account of the British Hieracia*. London.

[LOCKHART, J. G.] (1837). *Memoirs of the life of Sir Walter Scott, Bart.*, **3**: 144.
    (Sir Walter Scott noted the presence of tormentil on the mossy moors near Lerwick
    in 1814.)

LOUSLEY, J. E. (1968). 'A glabrous perennial *Sonchus* in Britain'. *Proceedings of
    the Botanical Society of the British Isles*, **7**: 151–157.

LOUSLEY, J. E. & KENT, D. H. (1981). *Docks and Knotweeds of the British Isles*.
    B.S.B.I. Handbook, no. 3.

LOW, G. See Anderson (1879).

LUSBY, P. S. (1984). 'The History and Taxonomy of the Shetland Taxon of
    Cerastium arcticum Lange'. (A thesis submitted for a B.Sc. degree in the
    University of Aberdeen.)

LYALL, G. (1996). 'Shetland plants in South London'. *BSBI News*, no. 72: 66.

McALLISTER, H. A. (1997). '*Vaccinium uliginosum* ssp. *cf. microphyllum*'.
    Botanical journal of Scotland, **49**: 274–275.

McCLINTOCK, D. & YEO, P. F. (1968). 'Geranium ibericum Cav., G. playtpetalum
    Fisch. & Mey. and G. × magnificum Hyl.'. *Proceedings of the Botanical Society of
    the British Isles*, **7**: 389.

MACKECHNIE, R. (1963). 'Unst'. *Proceedings of the Botanical Society of the
    British Isles*, **5**: 84–85. (An account of the BSBI field meeting, 1960, while based
    in Unst, and later in South Mainland.)

McKEE, J. (1995). 'A surprising sunflower'. *Shetland Times*, 29 September 1995: 13.

M'NAB, [J.] (1873). '*Remarks on* Juncus effusus spiralis *and the Varieties of Ferns*'.
    Transactions of the Botanical Society [of Edinburgh], **11**: 502–504.

MACPHERSON, P., DICKSON, J. H., ELLIS, R. G., KENT, D. H. & STACE, C. A.
    (1996). 'Plant status nomenclature'. *BSBI News*, no. 72: 13–16.

MALCOLM, D. (1992). *Shetland's wild flowers. A Photographic Guide*, ed. 2. Lerwick.

MANSON, T. M. Y. (1932a). *Mansons' Guide to Shetland*, ed. 1. Lerwick.

——— (1932b). *Mansons' Guide to Shetland*, ed. 2. Lerwick.

MARSHALL, E. S. (1895a). 'Cochlearia micacea Marshall in Shetland'. *Journal of Botany, British and foreign*, **33**: 152–153.

——— (1895b), *in* Linton, W. R. (ed.), *Botanical Exchange Club of the British Isles. Report for 1894*. (A note on scurvygrass from Unst on p. 435.)

——— (1910). 'William Hadden Beeby. (1849–1910)'. *Journal of Botany, British and foreign*, **48**: 121–123, plus portrait. (An obituary.)

MARSHALL, E. S. & SHOOLBRED, W. A. (1898). 'Notes of a tour in N. Scotland, 1897'. *Journal of Botany, British and foreign*, **36**: 166–177.

MARSHALL, J. G. S. (1968), ed. 'New wild Pilosella'. *Gardeners['] Chronicle*, 12 January 1968: 30.

MARTINS, C. (*c*.1848). 'Essai sur la végétation de l'archipel des Féröe, comparée a celle des Shetland et de l'Islande méridionale'. *Voyages de la Commission scientifique du Nord, en Scandinavie, en Laponie,au Spitzberg et aux* Féröe, **2**: 353–450.

MATHER, A. S. & SMITH, J. S. (1974). 'Beaches of Shetland'. Department of Geography, University of Aberdeen. (Commissioned by the Countryside Commission for Scotland.)

MATTHEWS, J. R. & ROGER, J. G. (1941). 'Variation in *Trientalis europaea* Linn.'. *Journal of Botany, British and foreign*, **79**: 80–83.

MEIKLE, R. D. (1984). *Willows and Poplars of Great Britain and Ireland*. B.S.B.I. Handbook, no. 4.

MEIKLEJOHN, M. F. M. (1959), *in* Wallace, E. C., 'Plant records'. *Proceedings of the Botanical Society of the British Isles*, **3**: 181–203.

MESSENGER, K. G. (1971). 'Notes on the distribution of selected plant species. Foula—August—1970'. *Brathay Exploration Group. Field Studies Report*, no. 11: 59–63.

——— (1974a). 'Records of vascular plants. Foula 1973'. *Brathay Exploration Group. Field Studies Report*, no. 27: 13–18.

——— (1974b). 'The genus Euphrasia in Foula, 1973'. *Brathay Exploration Group. Field Studies Report*, no. 27: 32–49.

MESSENGER, K. G. & URQUHART, J. G. (1959). 'Additions to the Flora of Foula (V.-C. 112)'. *Transactions and Proceedings of the Botanical Society of Edinburgh*, **37**: 276–278.

MILL, J. (1793). 'Parish of Dunrossness in Zetland', *in* Sinclair, J. (comp.), *The statistical account of Scotland*, **7**: 391–398. (References to rowan and willows on p. 392, but no indication as to their indigenousness or otherwise.)

MILNER, C. (1978). 'Shetland ecology surveyed'. *Geographical Magazine*, **50**: 730–736.

MITCHELL, A. (1974). *A field guide to the Trees of Britain and Northern Europe*. London.

MORTIMER, M. A. E. (1971). 'Mill Loch—Foula, Shetland'. *Brathay Exploration Group. Field Studies Group*, no. 11: 69–73.

142

———— (1973), ed. 'Field Studies Report no. 11: additions, amendments and comments'. *Brathay Exploration Group. Field Studies Report*, no. 21: 2–4.

MORTON, T. (2001). 'Hawkweed study needs climbing enthusiasts with nerves of steel'. *Shetland Times*, 10 August 2001: 4. (An account of the hawkweed study carried out in parts of the north and west of Shetland in 2001 by B. Vincent.)

MOSS, C. E. (1914). *The Cambridge British flora*, **2**. Cambridge. (On p. 45 there is an almost certainly erroneous reference to tea-leaved willow in Shetland.)

———— (1920). *The Cambridge British flora*, **3**. Cambridge.

MOUAT, T. & BARCLAY, J. (1793). 'Island and Parish of Unst, in Shetland', *in* Sinclair, J. (comp.), *The statistical account of Scotland*, **5**: 182–202.

MYKURA, W. (1976). *British Regional Geology. Orkney and Shetland.* Edinburgh.

NATURE CONSERVATION (SCOTLAND) ACT 2004. The Stationery Office.

NEILL, P. (1805). 'Remarks *made in a* Tour *thro' some of the* Shetland Islands *in* 1804'. *Scots Magazine*, **67**: 347–352, 431–435.

———— (1806). *A Tour through some of the Islands of Orkney and Shetland.* Edinburgh.

NELSON, E. C. (1995). 'Notes on some Shetland Islands specimens in the National Botanic Gardens, Glasnevin, Dublin (DBN)'. *Shetland Naturalist*, **1**: 101–108.

———— (2000). *Sea Beans and Nickar Nuts.* [B.S.B.I. Handbook, no. 10.]

NEUSTEIN, S. A. (1964). 'A review of pilot and trial plantations established by the Forestry Commission in Shetland'. *Scottish Forestry*, **18**: 199–211.

NEWBOLD, C. (1986). 'A Nature Conservation Assessment of the Flora of some Freshwater Shetland Lochs'. (A report to the Nature Conservancy Council, Peterborough.)

NEWMAN, E. (1841). 'The savin-leaved club-moss'. *The Phytologist: a popular botanical miscellany*, **1**: 33–36. (A record by T. Edmondston of alpine clubmoss from Unst appears on p. 34.)

———— (1844). *A History of British Ferns, and Allied Plants*, [ed.2]. London. (A record by T. Edmondston of Wilson's filmy-fern from a stream near Skaw, Unst, appears on p. 327.)

N[ICOLSON], J. R. (1977). 'Scalloway Notes'. *Shetland Times*, 21 October 1977. (Scarlet pimpernel is recorded from a Scalloway garden, on p. 18.)

NICOLSON, J. R. (1982). 'Scalloway notes'. *Shetland Times*, 25 June 1982. (Contains an item, 'Our native willows', on p. 18.)

NOLAN A. J., HULME, P. D. & WHEELER, D. (1994). 'The Status of *Calluna vulgaris* (L.) Hull on Fair Isle'. *Botanical Journal of Scotland*, **47**: 1–16.

NORDE, R. (2003). 'Mousa. NVC Survey 2002'. (A report to the Royal Society for the Protection of Birds.)

NORDHAGEN, R. (1966). 'Remarks on the serpentine-sorrel, Rumex acetosa subsp. serpentinicola (Rune) Nordhagen, and its distribution in Norway'. *Blyttia*, **24**: 286–294.

OCKENDON, D. J. & WALTERS, S. M. (1970). 'Studies in *Potentilla anserina* L.' *Watsonia*, **8**: 135–144.

'OLD WICK'. See Tudor (1883).

OSTENFELD, C. H. (1901). 'Phanerogamae and pteridophyta', *in* Warming, E. (ed.), *Botany of the Færöes*, pt. 1: 41–99. Copenhagen.

OSTENFELD, C. H. & GRÖNTVED, J. (1934). *The flora of Iceland and the Færoes.* Copenhagen.

PADMORE, P. A. (1957). 'The varieties of *Ranunculus flammula* L. and the status of *R. scoticus* E. S. Marshall and of *R. reptans* L.'.

PALMER, B. (2006), ed. *Richard Palmer. A Life in Letters.* Published by the author.

PALMER, R. C. (1957a), *in* Wallace, E. C. (comp.), 'Plant records'. *Proceedings of the Botanical Society of the British Isles*, **2**: 245–268.

—— (1957b), *in* Wallace, E. C. (comp.), 'Plant records'. *Proceedings of the Botanical Society of the British Isles*, **2**: 367–382.

—— (1959), *in* Wallace, E. C. (comp.), 'Plant records'. *Proceedings of the Botanical Society of the British Isles*, **3**: 291–300.

—— (1960), *in* Wallace, E. C. (comp.), 'Plant records'. *Proceedings of the Botanical Society of the British Isles*, **4**: 46–54.

—— (1962), *in* Wallace, E. C. (comp.), 'Plant records'. *Proceedings of the Botanical Society of the British Isles*, **4**: 419–433.

—— (1963a), *in* Wallace, E. C. (comp.), 'Plant records'. *Proceedings of the Botanical Society of the British Isles*, **5**: 28–41.

—— 1963b). 'Lerwick'. *Proceedings of the Botanical Society of the British Isles*, **5**: 81–83. (An account of the BSBI field meeting, 1960, while based in Lerwick.)

—— (1963c), *in* Wallace, E. C. (comp.), 'Plant records'. *Proceedings of the Botanical Society of the British Isles*, **5**: 125–143.

—— (1964), *in* Wallace, E. C. (comp.), 'Plant records'. *Proceedings of the Botanical Society of the British Isles*, **5**: 234–240.

—— (1966). 'Impressions of an English Botanist in Shetland'. *New Shetlander*, no. 76: 12–14.

—— (1967), *in* Wallace, E. C. (comp.), 'Plant records'. *Proceedings of the Botanical Society of the British Isles*, **7**: 27–32.

—— (1977). 'Some recent discoveries in Shetland'. *Watsonia*, **11**: 430.

—— (1992). 'White-flowered *Ranunculus acris*'. *B.S.B.I. News*, no. 60: 8.

PALMER, R. C. & SCOTT, W. (1957), *in* Wallace, E. C. (comp.), 'Plant records'. *Proceedings of the Botanical Society of the British Isles*, **2**: 245–268.

—— (1959), *in* Wallace, E. C. (comp.), 'Plant records'. *Proceedings of the Botanical Society of the British Isles*, **3**: 291–300.

—— (1962), *in* Wallace, E. C. (comp.), 'Plant records'. *Proceedings of the Botanical Society of the British Isles*, **4**: 419–433.

—— (1963), *in* Wallace, E. C. (comp.), 'Plant records'. *Proceedings of the Botanical Society of the British Isles*, **5**: 28–41.

—— (1965). 'Yet more additions to the flora of Fair Isle'. *Proceedings of the Botanical Society of the British Isles*, **6**: 43–45.

—— (1969). *A Check-list of the Flowering Plants and Ferns of the Shetland Islands.* Published by the authors.

—— (1995). 'A forgotten Shetland dandelion'. *Watsonia*, **20**: 279–281. (A short history and description of the Shetland dandelion (*Taraxacum geirhildae*).)

—— (1996). 'Two newly recognised Shetland species: *Taraxacum geirhildae* and *Hieracium spenceanum*'. *BSBI News*, no. 72: 68–69.

PARNELL, R. (1842). *The grasses of Scotland.* Edinburgh. (The first Scottish record of rough dog's-tail (from Bressay, where it was seen on two occasions) appears on p. 67.)

PEARSALL, [W. H.] (1935). '*Potamogeton polygonifolius* Pourr., var. *lancifolius* (Ch. & Schl.) Asch. et Gr., f. nov. *attenuatus* Pears.'. *Botanical Society and Exchange Club of the British Isles. Report for 1934*, **10**: 987–988.

PEDERSEN, A. (1966). '*Lathyrus maritimus* (L.) Fries i Danmark, en deling i to arter'. *Flora og Fauna*, **72**: 125–129.

PENNANT, T. (1784). *Arctic zoology*, **1**: xxx–xxxi. (Pennant contrasts the abundance of the remains of trees dug out of the peat in Orkney and Shetland with the scarcity of living trees there. His list of extant shrubs (p. xxxi) appears to be a general reference to both counties; there is no specific record from either Orkney or Shetland.)

PENNIE, I. (1985). 'Fair Isle Natural History Course 16th to 23rd June, 1984'. *Fair Isle Bird Observatory. Report for 1984*, 13–14.

PERRING, F. H. & SELL, P. D. (1968), eds. *Critical Supplement to the Atlas of the British Flora.* London.

PERRING, F. H. & WALTERS, S. M. (1962), eds. *Atlas of the British Flora.* London.

PERRY, R. (1948). *Shetland sanctuary.* (On p. 42 trailing azalea is recorded from sea-level in Bressay; an inexplicable error.)

PETTET, A. (1964). 'Studies on British pansies. II. The status of some intermediates between *Viola tricolor* L. and *V. arvensis* Murr.'. *Watsonia*, **6**: 51–69.

PILGER, R. (1937). 'Plantaginaceae', *in* Engler, A. (ed.), *Das Pflanzenreich*, **102**: 181.

PLOYEN, C. (1840). *Erindringer fra en Reise til Shetlandsöerne, Örkenöerne og Skotland i Sommeren 1839.* Copenhagen. (References to wild roses and honeysuckle at Fladdabister, South Mainland, on p. 80, and to angelica of Faeroese origin in a garden at Baltasound, Unst, on p. 292. English translation (by C. Spence): *Reminiscences of a voyage to Shetland, Orkney & Scotland in the summer of 1839* (Lerwick, 1894). Same references on p. 60, 232–233.)

POBEDIMOVA, E. (1968). 'Species novae generis Cochlearia L'. *Novitates systematicae plantarum vascularium*, [**5**]: 130–135.

POLAND, J. & CLEMENT, E. J. (2009). *The Vegetative Key to the British Flora.* Southampton.

PRENTICE, H. C. (1980). 'Variation in *Silene dioica* (L.) Clairv. : numerical analysis of populations from Scotland'. *Watsonia*, **13**: 11–26.

PRESTON, C. D. (1994). '*Juncus effusus* var. *spiralis* J. McNab in the Inner Hebrides'. *Watsonia*, **20**: 153–154.

——— (1995). *Pondweeds of Great Britain and Ireland.* B.S.B.I. Handbook, no. 8.

——— (1999). 'Some overlooked specimens of *Potamogeton × suecicus*'. *BSBI Scottish Newsletter*, no. 21: 14–18.

PRESTON, C. D., BAILEY, J. P. & HOLLINGSWORTH, P. M.(1998). 'A reassessment of the hybrid *Potamogeton × gessnacensis* G. Fisch. (*P. natans × P. polygonifolius*, Potamogetonaceae) in Britain'. *Watsonia*, **22**: 61–68.

PRESTON, C. D. & CROFT, J. M. (1997). *Aquatic plants in Britain and Ireland.* Colchester.

PRESTON, C. D., HOLLINGSWORTH, P. M. & GORNALL, R. J. (1999). 'The distribution and habitat of *Potamogeton* × *suecicus* K. Richt. (*P. filiformis* Pers. × *P. pectinatus* L.) in the British Isles'. *Watsonia*, **22**: 329–342.

PRESTON, C. D., PEARMAN, D. A. & DINES, T. D. (2002), eds. *New atlas of the British & Irish flora.* Oxford.

PRICE, W. R. (1929a). '*Betula pubescens* Ehrh. in Shetland'. *Journal of Botany, British and foreign*, **67**: 27–28. (A note on downy birch on a holm in a loch in North Roe, Northmavine.)

—— (1929b), *in* Druce, G. C. (ed.), 'New county and other records'. *Botanical Society and Exchange Club of the British Isles. Report for 1928*, 722–769. (A further note on downy birch in North Roe on p. 758. See Price (1929a).)

—— (1929c). 'Notes on the vegetation of Zetland, 1928'. *Botanical Society and Exchange Club of the British Isles. Report for 1928*, 770–781.

PRITCHARD, N. M. (1957). 'Notes on the flora of Fair Isle '. *Proceedings of the Botanical Society of the British Isles*, **2**: 218–225.

—— (1960). '*Gentianella* in Britain. II. *Gentianella septentrionalis* (Druce) E. F. Warb'. *Watsonia*, **4**: 218–237.

PROCTOR, J. (1971). 'The plant ecology of serpentine. II. Plant response to serpentine soils'. *Journal of Ecology*, **59**: 397–410.

PROCTOR, J. & WOODELL, S. R. J. (1971). 'The plant ecology of serpentine. I. Serpentine vegetation of England and Scotland'. *Journal of Ecology*, **59**: 375–395.

PUGSLEY, H. W. (1930). 'A Revision of the British Euphrasiæ'. *Journal of the Linnean Society of London. Botany*, **48**: 467–544, plus twelve plates.

—— (1933a). 'Notes on British Euphrasias. — III'. *Journal of Botany, British and foreign*, **71**: 83–90.

—— (1933b). 'The Euphrasias of Iceland and the Faroes'. *Journal of Botany, British and foreign*, **71**: 303–309.

—— (1935). 'On some Marsh Orchids'. *Journal of the Linnean Society of London. Botany*, **49**: 553–592.

—— (1948). 'A prodromus of the British Hieracia'. *Journal of the Linnean Society of London. Botany*, **54**.

RAATIKAINEN, T. (1961). 'Studies on Finnish populations of Sorbus aucuparia L.' *Archivum Societatis Zoologicae Botanicae Fennicae 'Vanamo'*, **15**: 64–82.

RAEBURN, H. (1888). 'The Summer Birds of Shetland, with Notes on their Distribution, Nesting, and Numbers'. *Proceedings of the Royal Physical Society of Edinburgh*, **9**: 542–562. (Raeburn erroneously lists ash as one of the natural trees found on some of the 'islets in lochs and inland cliffs' of Shetland.

—— (1891). 'The birds of Papa Stour, with an account of the Lyra Skerry'. *The Zoologist*, **15** (third series): 126–135.

RANDALL, R. E. (1974). '*Rorippa islandica* (Oeder) Borbás *sensu stricto* in the British Isles'. *Watsonia*, **10**: 80–82.

—— (1977). 'The past and present status and distribution of Sea Pea, *Lathyrus maritimus* Willd., in the British Isles'. *Watsonia*, **11**: 247–251.

RASMUSSEN, R. (1952). *Føroya flora*, ed. 2. Torshavn.

RIBBONS, B. W. (1962), *in* Wallace, E. C. (comp.), 'Plant records'. *Proceedings of the Botanical Society of the British Isles*, **4**: 419–433.

RICH, T. C. G. (1991). *Crucifers of Great Britain and Ireland*. B.S.B.I. Handbook, no.6.

RICH, T. C. G. & DALBY, D. H. (1996). 'The Status and Distribution of Mountain Scurvygrass (*Cochlearia micacea* Marshall) in Scotland, with Ecological Notes'. *Botanical Journal of Scotland*, **48**: 187–198.

RICH, T. C. G. & JERMY, A. C. (1998). *Plant crib 1998*. London.

RICHARDS, A. J. (1972). 'The *Taraxacum* flora of the British Isles'. *Watsonia*, **9** (supplement).

RICHARDS, A. J. & HAWORTH, C. C. (1984). 'Further new species of *Taraxacum* from the British Isles'. *Watsonia*, **15**: 85–94.

RIDDIFORD, N. J. (2010). 'Making Waves. Fair Isle Marine Environment and Tourism Initiative', newsletter no. 7, pt. 1. <http://www.fimeti.org.uk> Last accessed 18 February 2011. (An account of oysterplant in Fair Isle on p. 14–15.)

RIDDINGTON, R. (1999). 'Biological records in Shetland. An audit of existing information.' (A report to the Shetland Biological Records Centre.)

———— (2000). 'Flood prompts plant rescue'. *Shetland Times*, 29 September 2000: 26. (A note of the damage sustained by an intermediate juniper bush at Levenwick, South Mainland.)

RIDLEY, H. N. (1884). 'Shetland Plants'. *Journal of Botany, British and foreign*, **22**: 301. (Refers to sea-holly being found in 1884 by W. E. Smith at Fitful Head, South Mainland.)

ROBERTS, R. H. (1964). '*Mimulus* hybrids in Britain'. *Watsonia*, **6**: 70–75.

———— (1968). 'The hybrids of *Mimulus cupreus*'. *Watsonia*, **6**: 371–376.

ROBERTS, R. H. & STIRLING, A. McG. (1974). 'Asplenium cuneifolium Viv. in Scotland'. *Fern Gazette*, **11**: 7–14.

ROBERTSON, J. 1769 (a). 'Observations made in a Tour thro' the Islands of Orkney and Shetland in the year 1769'. MS in Shetland Museum and Archives, Hay's Dock, Lerwick, Shetland.

———— 1769 (b) . 'Flora and Fauna of Shetland'. Henderson & Dickson (1994) record that this MS was found in the Signet Library, Parliament Square, Edinburgh.

———— *c*.1771. 'Flora and Fauna of the Islands of Scotland'. Henderson & Dickson (1994) record that this MS was found in the Signet Library, Parliament Square, Edinburgh.

RONNIGER, K. (1924). 'Contributions to the knowledge of the genus Thymus'. *Botanical Society and Exchange Club of the British Isles. Report for 1923*, **7**: 226–239.

———— (1928). 'The distribution of Thymus in Britain'. *Botanical Society and Exchange Club of the British Isles. Report for 1927*, **8**: 509–517.

ROPER-LINDSAY, J. & SAY, A. M. (1986). 'Plant communities of the Shetland Islands'. *Journal of Ecology*, **74**: 1013–1030.

RUNE, O. (1957). 'De serpentinicola elementen i Fennoskandiens flora'. *Svensk Botanisk Tidskrift*, **51**: 43–105.

SALAMAN, R. N. (1926). *Potato varieties*. Cambridge.

SAXBY, C. F. A. (1903). *Edmondston's Flora of Shetland*. Edinburgh. (The so-called 'second edition' of Edmondston (1845). A 'third edition' exists but it is merely a re-issue of the present work.)

SAXBY, H. L. (1874). *The birds of Shetland*. Edinburgh. (On p. 85 there is a reference to scarlet pimpernel in the garden of 'Halligarth', Baltasound, Unst.)

SAXBY, J. (1947). 'A hobbaviti'. *Shetland folk book*, **1**: 58–62. (References to wild roses and honeysuckle at the head of Gloup Voe, Yell, on p. 59.)

SAXBY, J. M. E. (1909). 'Shetland names for animals, etc. II'. *Old-lore Miscellany of Orkney Shetland Caithness and Sutherland*, **2**: 235–237.

——— (1934). 'Wild flowers of Hialtland', *in* Saxby, J. M. E., *Threads from a tangled skein*, 346–353. Lerwick.

SCHMID, B. (1983). 'Notes on the nomenclature and taxonomy of the *Carex flava* group in Europe'. *Watsonia*, **14**: 309–319.

SCOTT, L. G. (1958). 'A trip to the Ve Skerries'. *New Shetlander*, no. 47: 12–14.

SCOTT, W. (1956), *in* Wallace, E. C. (comp.), 'Plant records'. *Proceedings of the Botanical Society of the British Isles*, **2**: 138–150.

——— (1957), *in* Wallace, E. C. (comp.), 'Plant records'. *Proceedings of the Botanical Society of the British Isles*, **2**: 245–268.

——— (1958), *in* Wallace, E. C. (comp.), 'Plant records'. *Proceedings of the Botanical Society of the British Isles*, **3**: 53–65.

——— (1959a), *in* Wallace, E. C. (comp.), 'Plant records'. *Proceedings of the Botanical Society of the British Isles*, **3**: 181–203.

——— (1959b). 'Notes on the flora of Shetland: Unst'. *New Shetlander*, no. 49: 27–29.

——— (1959c), *in* Wallace, E. C. (comp.), 'Plant records'. *Proceedings of the Botanical Society of the British Isles*, **3**: 291–300.

——— (1959d). 'Notes on the flora of Shetland'. *New Shetlander*, no. 51: 23–24.

——— (1960a). 'Notes on the Flora of Shetland'. *New Shetlander*, no. 54: 20–21.

——— (1960b), *in* Wallace, E. C. (comp.), 'Plant records'. *Proceedings of the Botanical Society of the British Isles*, **4**: 46–54.

——— (1961). 'Notes on the Flora of Shetland'. *New Shetlander*, no. 57: 25–27.

——— (1962a). 'Notes on the flora of Shetland'. *New Shetlander*, no. 60: 14–15.

[SCOTT, W.] (1962b). 'Profile from the past. No. XVI. Thomas Edmondston'. *New Shetlander*, no. 62:7–9.

SCOTT, W. (1963a), *in* Wallace, E. C. (comp.), 'Plant records'. *Proceedings of the Botanical Society of the British Isles*, **5**: 28–41.

——— (1963b). 'Grass'. *Shetland Times*, 15 March 1963: 3. (A note on tussac-grass.)

——— (1963c). 'Notes on the Flora of Shetland'. *New Shetlander*, no. 65:16–18.

——— (1963d), *in* Wallace, E. C. (comp.), 'Plant records'. *Proceedings of the Botanical Society of the British Isles*, **5**: 125–143.

——— (1963e). 'Notes on the flora of Shetland'. *New Shetlander*, no. 67: 32–34. (Brief notes on some of the plants of Fair Isle.)

——— (1964a), *in* Wallace, E. C. (comp.), 'Plant records'. *Proceedings of the Botanical Society of the British Isles*, **5**: 234–240.

——— (1964b). 'Notes on the flora of Shetland'. *New Shetlander*, no. 69: 9–10.

—— (1964c). 'Notes on the Flora of Shetland'. *New Shetlander*, no. 71: 30–32.

—— (1965). 'Notes on the Flora of Shetland'. *New Shetlander*, no. 73: 26–27.

—— (1966a). 'Notes on the flora of Shetland'. *New Shetlander*, no. 77: 29–30.

—— (1966b), *in* Wallace, E. C. (comp.), 'Plant records'. *Proceedings of the Botanical Society of the British Isles*, **6**: 235–246.

—— (1966c). 'W. H. Beeby and the Botany of Shetland'. *New Shetlander*, no. 79: 11–14.

—— (1967). 'Notes on the Flora of Shetland'. *New Shetlander*, no. 82: 16–17. (Some notes on the Shetland Mouse-ear-hawkweed.)

—— (1968a). 'Recent plant discoveries in Shetland'. *New Shetlander*, no. 84: 11–12.

—— (1968b). 'Pilosella flagellaris (Willd.) Sell & C. West subsp. bicapitata Sell & C. West—in Zetland'. *Proceedings of the Botanical Society of the British Isles*, **7**: 192–193.

—— (1968c). 'Notes on the Flora of Shetland'. *New Shetlander*, no. 86: 19–21.

—— (1969). 'Notes on the flora of Shetland'. *New Shetlander*, no. 91: 17–19.

—— (1970a). '1st–8th August'. *Watsonia*, **8**: 191–193. (An account of the BSBI field meeting, 1960, based in Lerwick.)

—— (1970b). 'Notes on the Flora of Shetland'. *New Shetlander*, no. 94: 28–29. (Notes on the plants of the Ve Skerries and the Muckle Ossa.)

—— (1972). 'A check-list of the flora of Fair Isle'. *Fair Isle Bird Observatory. Report for 1971*: 57–70.

—— (2011). *Some aspects of the botany of the Shetland Islands.* (In preparation.)

SCOTT, W., HARVEY, P., RIDDINGTON, R. & FISHER, M. (2002). *Rare plants of Shetland.* [Lerwick.]

SCOTT, W. & PALMER, R. [C.] (1987). *The flowering plants and ferns of the Shetland Islands.* Lerwick.

—— (1957), *in* Wallace, E. C. (comp.), 'Plant records'. *Proceedings of the Botanical Society of the British Isles*, **2**: 245–268.

—— (1958), *in* Wallace, E. C. (comp.), 'Plant records'. *Proceedings of the Botanical Society of the British Isles*, **3**: 53–65.

—— (1963), *in* Wallace, E. C. (comp.), 'Plant records'. *Proceedings of the Botanical Society of the British Isles*, **5**: 28–41.

—— (1980). 'Flowering Plants and Ferns', *in* Berry, R. J. & Johnston, J. L., *The Natural History of Shetland*, 282–303. London. (New Naturalist series, no. 64.)

—— (1995). 'A new Shetland *Hieracium* of the Section *Alpestria* [Fries] F. N. Williams'. *Watsonia*, **20**: 282–284. (A description of Spence's hawkweed, a species of West Mainland.)

—— (1999). 'Flowering plants and ferns', *in* Johnston, J. L., *A Naturalist's Shetland*, 385–409. London.

SCOTTISH AGRICULTURAL COLLEGES (1976). 'Seed mixtures for Scotland'. Publication no. 12.

SCOTTISH NATURAL HERITAGE (1997). 'Shetland Loch Inventory 1997'.

—— [2009]. 'The Story of Keen of Hamar National Nature Reserve'.

—— [2009]. 'The Proposals for Keen of Hamar National Nature Reserve'.

SEEMANN, B. (1853). *Narrative of the voyage of H.M.S. Herald during the years 1845–1851*, **1**: 65–69. London. (An account of the accidental death of Thomas Edmondston in South America in 1846.)

SELL, P. [D.] & MURRELL, G. (2006). *Flora of Great Britain and Ireland*, **4**. Cambridge. (Shetland references under *Taraxacum*, *Pilosella*, and *Hieracium*.)

SELL, P. D. & WEST, C. (1956). 'A Shetland "Endemic" *Hieracium*'. *Proceedings of the Botanical Society of the British Isles*, **2**: 79. (A note on the sparse-leaved hawkweed.)

―――― (1962). 'Notes on British Hieracia. II. The species of the Orkney Islands'. *Watsonia*, **5**: 215–223.

―――― (1965). 'A revision of the British species of *Hieracium* Section *Alpestria* [Fries] F. N. Williams'. *Watsonia*, **6**: 85–105.

―――― 1968). 'Hieracium L.', and 'Pilosella Hill', *in* Perring, F. H. & Sell, P. D. (eds.), *Critical Supplement to the Atlas of the British flora*, 75–134. London.

SETON, A. (1806). '*Letter ... containing Observations on the state of the Shetland Islands, and on the means of their improvement* ', *in* Neill, P., *A Tour through some of the Islands of Orkney and Shetland*, 173–181.

SHETLAND ISLANDS COUNCIL (2009). *Shetland in Statistics*, no. 36: 21.

SHEWRY, P. R. & PETERSON, P. J. (1975). 'Calcium and magnesium in plants and soil from a serpentine area on Unst, Shetland'. *Journal of Applied Ecology*, **12**; 381–391.

SHIRREFF, J. (1813). *General view of the Agriculture of Shetland*. Edinburgh. (In addition to a handful of references to some common species, there is an interesting mention of a plant (referred to as 'tares or grey pease' in a '*Letter from a Gentleman in Shetland* ' (appendix, p. 20). Shirreff later (1814, appendix, p. 20) points out that the plant in question was *Vicia sepium* (bush vetch).)

―――― (1814). *General view of the Agriculture of the Shetland Islands*. Edinburgh.

SHIVAS, M. G. (1962). 'The Polypodium vulgare complex'. *British Fern Gazette*, **9**: 65–70.

SIBBALD, R. (1711), publ. *The Description of the Isles of Orknay and Zetland*. Edinburgh.

SILVERSIDE, A. J. (1984). 'The subspecies of Ranunculus flammula'. *B.S.B.I. Scottish newsletter*, no. 6: 4–7.

―――― (1994). '*Mimulus*: 180 years of confusion' *in* Perry, A. R. & Ellis, R. G. (eds.), *The Common Ground of Wild and Cultivated Plants*, 59–64. Cardiff. Part of Table 3 was inadvertently omitted from this article. The table appears in its entirety on p. 56 of *BSBI News*, no. 77 (1997).

―――― (1998). 'Mimulus L.'. >http://www-biol.paisley.ac.uk/bioref/ Plantae_Mimulus/Mimulus html< Last accessed 16 June 2007.

S[IM], J. (1878). 'Where to go for a holiday—Shetland'. *Daily Free Press*. Aberdeen. Published in seven issues from 9 July to 24 September. Reprinted in the *Shetland Times* from 3 August 1878 onwards. Fuller details appear in the bibliography in Scott & Palmer (1987).

SIMPSON, N. D. (1960). *A Bibliographical Index of the British Flora*, 386–387. Privately printed.

SINCLAIR, C. (1840). *Shetland and the Shetlanders*. Edinburgh. (On p. 130 the authoress notes a large rhubarb plant near Fort Charlotte, Lerwick. Catherine Sinclair was the daughter of Sir John Sinclair, the creator and tireless compiler of the monumental statistical account of Scotland. See next entry.)

SINCLAIR, J. (1791–1799). *The statistical account of Scotland*. Twenty-one volumes. Edinburgh. (See Tait (1925) on the accessibility of the Shetland accounts.)

—— (1795). *General View of the Agriculture of the Northern Counties and Islands of Scotland*, 250–251, and appendix no. 4, 'On Shetland sheep', by Thomas Johnston, p. 20–33. London.

SLEEP, A. (1980). 'On the reported occurrence of Asplenium cuneifolium and A. adiantum-nigrum in the British Isles'. *Fern Gazette*, **12**: 103–107.

SLEEP, A., ROBERTS, R. H., SOUTER, J. I. & STIRLING, A. McG. (1978). 'Further investigations on Asplenium cuneifolium in the British Isles'. *Fern Gazette*, **11**: 345–348.

SLINGSBY, D. R. (1978). '*The Keen of Hamar, Shetland NNR and SSSI: A survey of vegetation and of the numbers and distribution of some of the rarer species*'. (A report to the Nature Conservancy Council.)

—— (1979). '*The Keen of Hamar, Shetland NNR and SSSI. A survey and monitoring scheme*'. (A report to the Nature Conservancy Council.)

—— (1981a). 'Britain's most northerly desert'. *Shetland Life*, no. 12: 30–31.

—— (1981b). 'The Keen of Hamar, Shetland: A General Survey and a Census of some of the Rarer Plant Taxa'. *Transactions of the Botanical Society of Edinburgh*, **43**: 297–306.

—— (1982). 'The Vegetation of Crussa Field, Nikkavord and the Heogs, Unst, Shetland, 1982'. (A report to the Nature Conservancy Council.)

—— (1991a). 'The Keen of Hamar—a twenty-one year study'. *Shetland Naturalist*, **1**: 1–12.

—— (1991b). 'Monitoring the Keen of Hamar'. (A report to the Nature Conservancy Council.)

—— (1992a). 'The Keen of Hamar, Shetland—A Long-Term Site-Specific Study of a Classic Serpentine Site', *in* Baker, A. J. M., Proctor, J. & Reeves, R. D. (eds.), *The Vegetation of Ultramafic (Serpentine) Soils*, 235–241. Andover, Hampshire. (Proceedings of the first international conference on serpentine ecology held at the University of California, Davis, 19–22 June 1991.)

—— (1992b). 'Botanical monitoring of the Keen of Hamar'. (A report to Scottish Natural Heritage.)

—— (1994). 'The status of populations of rare plant taxa on the Keen of Hamar 1978–1993' (A report to Scottish Natural Heritage.)

SLINGSBY, D. R. & CARTER, S. (1986a). 'A comparative study of the morphology and ecology of *Cerastium arcticum* in Shetland and Faroe'. *British Ecological Society. Bulletin*, **17**: 25–28.

—— (1986b). 'The Ecological Effects of Eutrophication on the Keen of Hamar SSSI, Shetland'. (A report to the Nature Conservancy Council.)

—— (1987). 'Experimental Study into the Restoration of Serpentine Debris Habitat damaged by Eutrophication'. (A report to the Nature Conservancy Council.)

SLINGSBY, D. R., CARTER, S., KING, R. J. & JENNER, C. (1983). 'The Serpentine Vegetation of Fetlar and Haaf Gruney'. (A report to the Nature Conservancy Council.)

SLINGSBY, D. R., HOPKINS, J., CARTER, S., DALRYMPLE, S. & SLINGSBY, A. (2010). 'Change and Stability. Monitoring the Keen of Hamar: 1978–2006'. >http://www.snh.org.uk/pdfs/publications/nnr/Keen_of_Hamar_NNR_Change_and_Stability.pdf< Last accessed 18 February 2011. 2010. (Covers many years of botanical surveying on the Keen of Hamar, Unst.)

SLINGSBY, D. R., PROCTOR, J. & CARTER, S. P. (2001). 'Stability and change in ultramafic fellfield vegetation at the Keen of Hamar, Shetland, Scotland'. *Plant Ecology*, **152**: 157–165.

SLINGSBY, D. R., ROGERS, L. C. F. & KENDALL, D. M. (1987). 'The Crussa Field/Heogs PSSSI, Unst, Shetland'. (A report to the Nature Conservancy Council.)

SMITH, P.M. (1973). 'Observations on some critical Bromegrasses'. *Watsonia*, **9**: 319–332.

——— (1975). ' *Endymion* Dumort.', *in* Stace, C. A. (ed.), *Hybridization and the Flora of the British Isles*, 460. London.

SMITH, R. J. (1999). 'Walter Scott MBE and the "Shetland *Dryopteris*"'. *Pteridologist*, **3**: 65–66.

SNOGERUP, B. (1982). '*Odontites litoralis* Fries subsp. *litoralis* in the British Isles'. *Watsonia*, **14**: 35–39.

SOUTH NESTING PUBLIC SCHOOL (1963), *in* Wallace, E. C. (comp.), 'Plant records'. *Proceedings of the Botanical Society of the British Isles*, **5**: 125–143. (A reference to biting stonecrop appears on p. 132.)

SPENCE, D. H. N. (1956). 'Studies on the vegetation of Shetland'. (A thesis submitted for a Ph.D. degree in the University of Glasgow.)

——— (1957a), *in* Wallace, E. C. (comp.), 'Plant records'. *Proceedings of the Botanical Society of the British Isles*, **2**: 245–268.

——— (1957b). 'Studies on the vegetation of Shetland. I. The serpentine debris vegetation in Unst'. *Journal of Ecology*, **45**: 917–945.

——— (1958). 'The Flora of Unst, Shetland, in relation to the Geology'. *Transactions and Proceedings of the Botanical Society of Edinburgh*, **37**: 163–173.

——— (1959). 'Studies on the vegetation of Shetland. II. Reasons for the restriction of the exclusive pioneers to serpentine debris'. *Journal of Ecology*, **47**: 641–649.

——— (1960). 'Studies on the vegetation of Shetland. III. Scrub in Shetland and in South Uist, Outer Hebrides'. *Journal of Ecology*, **48**: 73–95.

——— (1970). 'Scottish serpentine vegetation'. *Oikos*, **21**: 22–31.

——— (1974). 'Subarctic debris and scrub vegetation of Shetland', *in* Goodier, R. (ed.), *The natural environment of Shetland*, 73–88. Nature Conservancy Council, Edinburgh.

——— (1979). *Shetland's living landscape*. Sandwick, Shetland.

——— (1980). 'Vegetation', *in* Berry, R. J. & Johnston, J. L., *The Natural History of Shetland*, 73–91. London. (New Naturalist series, no. 64.)

SPENCE, D. H. N. & MILLAR, E. A. (1963). 'An experimental study of the infertility of a Shetland serpentine soil'. *Journal of Ecology*, **51**: 333–343.

SPENCE, M. (1914). *Flora Orcadensis*. Kirkwall.

STACE, C. A. (1975a), ed. *Hybridization and the Flora of the British Isles*. London.

—— (1975b). '*Epilobium* L.', *in* Stace, C. A. (ed.), *Hybridization and the Flora of the British Isles*. London.

—— (1997). *New flora of the British* Isles, ed. 2. Cambridge.

—— (2010). *New flora of the British Isles*, ed. 3. Cambridge.

STEPPANOVA, R. (2004). *The Impossible Garden*. Lerwick.

STEWART, A., PEARMAN, D. A. & PRESTON, C. D. (1994), comps. *Scarce Plants in Britain*. Peterborough.

STEWART, G. G. (1962). 'Kergord Plantations, Shetland'. *Forestry*, **35**: 35–56.

[STEWART, W.] (1654). 'Nova descriptio Schetlandiae', *in* Blaeu, W. & J. (eds.), *Theatrum orbis terrarum, sive Atlas novus in quo tabulae et descriptiones omnium regionum*, **5**: 147–149. Amsterdam. (Contains a few references to heather. Irvine (2006) provides a translation by I. Cunningham of Stewart's description, and explains that it was written *c*.1645.)

STEWART, W. S. & BANNISTER, P. (1974). 'Dark Respiration Rates in *Vaccinium* spp. in Relation to Altitude'. *Flora*, **163**: 415–421.

STUFFINS, K. A. (1983). 'The growth of Cerastium arcticum ssp. edmondstonii (H. C. Watson) Á. and D. Löve and other Shetland plants in three prepared soils'. (A thesis submitted for a B.Sc. degree in Royal Holloway College, University of London.)

STYLES, B. T. (1962). 'The taxonomy of *Polygonum aviculare* and its allies in Britain'. *Watsonia*, **5**: 177–214.

SUMMERHAYES, V. S. (1968). *Wild orchids of Britain*, ed. 2. London. (New Naturalist series, no. 19.)

SYME, J. T. B. (1866), ed. *English Botany; or, coloured figures of British plants*, ed. 3, **5**:185–186. London.

TAIT, E. S. R. (1925), ed. *The statistical account of Shetland 1791–1799*. Lerwick. (Tait brings together, for the first time in a single volume, all the Shetland parish accounts which had hitherto appeared in no particular order in Sir John Sinclair's multi-volume compilation, *The statistical account of Scotland*.

TAIT, R. W. (1932). 'The flora of Shetland', *in* Manson, T. M. Y. (comp.), *Mansons' Guide to Shetland*, ed. 1: 80–87. Lerwick.

—— (1947). 'Some Shetland plant names'. *Shetland folk book*, **1** : 73–88.

TASCHEREAU, P. M. (1977). '*Atriplex praecox* Hülphers: a species new to the British Isles'. *Watsonia*, **11**: 195–198.

—— (1985). 'Taxonomy of *Atriplex* species indigenous to the British Isles'. *Watsonia*, **15**: 183–209.

—— (1986). 'Hybridization in the genus *Atriplex* section *Teutliopsis* (Chenopodiaceae)'. *Watsonia*, **16**: 153–162.

—— (1988). 'Taxonomy, morphology and distribution of *Atriplex* hybrids in the British Isles'. *Watsonia*, **17**: 247–264.

TATE, R. (1865). 'Alpine Plantain in Shetland'. *Hardwicke's Science-Gossip*, [**1**]: 283.

—— (1866). 'Upon the flora of the Shetland Isles'. *Journal of Botany, British and foreign*, **4**: 2–15. (See also Watson (1866) and Carruthers (1866). Tate was the first commentator on Shetland botany after Edmondston.)

THORNE, R. (1977). *Fetlar. Some facts and stories.* Probably published by the author.

THURLOW, G. & FOWLER, J. A. (1986). 'Relationship between number of higher plant species and island area on the Yell islands'. *Ecological studies in the maritime approaches to the Shetland oil terminal. 1984–1985*: 57–72. Leicester Polytechnic.

TOWNSEND, F. (1897). 'Monograph of the British Species of Euphrasia'. *Journal of Botany, British and foreign*, **35**; 321–336, 395–406, 417–426, 465–477, plus seven plates.

TRAIL, J. W. H. (1902). 'A new form of *Euphrasia curta,* Fr.'. *Annals of Scottish Natural History*, [**11**]: 177–178.

—— (1906). 'The flora of Fair Isle'. *Annals of Scottish Natural History*, [**15**]: 165–170.

TRAILL, T. S. (1806). '*Observations, chiefly mineralogical, on the Shetland Islands, made in the course of a Tour through those Islands in* 1803', *in* Neill, P., *A Tour through some of the Islands of Orkney and Shetland*, 157–173. Edinburgh. (On p. 161 there is a reference to bearberry in Muckle Roe, a shrub which still occurs in some parts of the island.)

TRAVIS, W. G. & WHELDON, J. A. (1912). 'A new variety of *Parnassia palustris*'. *Journal of Botany, British and foreign*, **50**: 254–257.

TRIST, P. J. O. & SELL, P. D. (1988). 'Two subspecies of *Molinia caerulea* (L.) Moench in the British Isles'. *Watsonia*, **17**: 153–157.

TUBBS, C. R. (1995). 'The Meadows in the Sea'. *British Wildlife*, **6**: 351–355. (An account of eelgrass in Britain and of the nationwide disease in the 1930s which decimated so many of its colonies including, it would appear, a few in Shetland.)

[TUDOR, J. R.] (1878). 'Rambling and angling notes from Shetland. The Walls District'. *The Field*, 13 July 1878: 67–68. (Written under the pseudonym 'Old Wick'; contains a reference to eelgrass in the Marlee Loch, West Mainland, on p. 68.)

TUDOR, J. R. (1883). *The Orkneys and Shetland; their Past and Present State.* London. (A reference to juniper in Fair Isle appears on p. 442.)

TURRILL, W. B. (1929). 'The flora of Foula'. *Botanical Society and Exchange Club of the British Isles. Report for 1928*, **8**: 838–850. (Largely based on J. Gladstone's visit to Foula in 1928.)

TUTIN, T. G. (1980). *Umbellifers of the British Isles.* B.S.B.I. Handbook, no. 2.

TYLER, G. (2003). 'Cape Pond Weed at the Vaadal Reservoir'. *Fair Isle Times*, 5 June 2003: [3].

URQUHART, J. T. (1824). 'On the Preparation of the Zostera or Sea-grass, in Orkney'. *Prize-essays and Transactions of the Highland Society of Scotland*, **6**: 588–593.

VENABLES, L. S. V. (1949). 'Kergord plantations'. *New Shetlander*, no. 16: 34–35.

VENABLES, L. S. V. & U. M. (1948). 'A Shetland bird population: Kergord plantations'. *Journal of Animal* Ecology, **17**: 66–74.

—— (1955). 'Average Flowering Dates of Some Shetland Plants', *in* Venables, L. S. V. & U. M., *Birds and mammals of Shetland*, 369–371. Edinburgh.

—— (1963), *in* Wallace, E. C. (comp.), 'Plant records'. *Proceedings of the Botanical Society of the British Isles*, **5**: 28–41.

VENABLES, U.[M.] (1952). *Tempestuous Eden*. London. (A short list of the trees and shrubs at Kergord Plantations appears on p. 34–41.)

—— (1956). *Life in Shetland*. (On p. 59 there is a reference to salmonberry at Tresta, Central Mainland, where this invasive North American alien persists to the present day.)

VINCENT, B. (2001). 'A Survey Of Shetland Hawkweeds (Hieracia) In The Northwest Mainland'. (A report to Scottish Natural Heritage, and a useful account which covers in large measure the more inaccessible sites of the area.)

WALKER, S. & JERMY, A. C. (1964). '*Dryopteris assimilis* S. Walker in Britain'. *British Fern Gazette*, **9**: 137–140.

WALTERS, S. M. (1949). '*Aphanes microcarpa* (Boiss. et Reut.) Rothm. in Britain'. *Watsonia*, **1**: 163–169.

—— (1953). '*Montia fontana* L.'. *Watsonia*, **3**: 1–6.

WALTON, D. W. H. (1982). 'Tussac grass'. *B.S.B.I. News*, no. 32: 12.

WARMING, E. (1901–1908), ed. *Botany of the Færöes*, pts. 1–3. Copenhagen. Issued in three parts (also called volumes): pt.1 (1901), pt. 2 (1903), pt. 3 (1908).

—— (1908). 'Field-notes on the biology of some of the flowers of the Færöes', *in* Warming, E. (ed.), *Botany of the Færöes*, pt. 3: 1055–1065.

WARREN, A. & HARRISON, C. M. (1974). 'A nature conservation plan for Shetland'. Discussion Papers in Conservation, no. 7. University College, London.

WATSON, H. C. (1843a). '*Note on the supposed new British Cerastium*'. *The Phytologist: a popular botanical miscellany*, **1**: 586–587.

—— (1843b). '*Note on the Cerastium latifolium of the Linnean herbarium*'. *The Phytologist: a popular botanical miscellany*, **1**: 717–718.

—— (1845). '*On the Cerastium latifolium* (Linn.) *var. Edmondstonii* (Lond. Cat.); *and on the Seeds of Cerastium latifolium and C. alpinum*'. *The Phytologist: a popular botanical miscellany*, **2**: 93–94.

—— (1849). *Cybele Britannica*, **2**. London. (References to wintergreen appear on p. 159–160.)

—— (1860). *Part first of a Supplement to the Cybele Britannica*. London. (The name *Cerastium nigrescens* is validly published on p. 81.)

—— (1866). 'Corrections in the Shetland flora'. *Journal of Botany, British and foreign*, **4**: 348–351. (Comments on Tate (1866).)

—— (1873, 1874). *Topographical Botany*, 'part first' (1873), 'part second' (1874). Printed for private distribution.

—— (1883). *Topographical Botany*, ed. 2. London.

WEST, W. (1912). 'Notes on the flora of Shetland, with some ecological observations'. *Journal of Botany, British and foreign*, **50**: 265–275, 297–306.

WETTSTEIN, R. VON (1896). *Monographie der Gattung Euphrasia*. Leipzig. (On p. 139–140 the Foula eyebright, which Beeby collected on Hamnafield (and from elsewhere in Shetland), is described for the first time.)

WHELDON, J. A. & TRAVIS, W. G. (1913). 'Parnassia palustris var. condensata'. *Journal of Botany, British and foreign*, **51**: 85–89.

WHITE, A. (1855). 'Note on the Natural History of Shetland'. *Proceedings of the Linnean Society of London*, 2: 157–158. (On p. 158 there is an erroneous reference to pyramidal bugle in Shetland.)

WHITE, P. 1880. 'Shetland Flora. Localities of Plants in Shetland arranged Alphabetically'. (Despite the promising title, this MS (in the possession of W. Scott, 'New Easterhoull', Castle Street, Scalloway, Shetland) is basically a repeat (without acknowledgment) of names, localities, and comments from T. Edmondston (1845).)

—— (1883). 'Notes on the flora of Shetland', *in* Tudor, J. R., *The Orkneys and Shetland; their Past and Present State*, 422–428. London.

WHITEHEAD, F. H. (1956). 'Taxonomic studies in the genus *Cerastium*. II. Cerastium subtetrandrum (Lange) Murb.'. *Watsonia*, 3: 324–326.

WHITELAW, C. & KIRKPATRICK, A. H. (1997). 'Heather Moorland Loss on the Northern Islands of Orkney'. *Botanical Journal of Scotland*, **49**: 57–65.

WHITTINGTON, G. & EDWARDS, K. J. (1993). 'Vegetation change on Papa Stour, Shetland, Scotland: a response to coastal evolution and human interference?'. *The Holocene*, 3: 54–62.

WIGGINTON, M. J. (1999). *British Red Data Books. I. Vascular plants*, ed.3. Peterborough.

WIGGINTON, M. J. & GRAHAM, G. G. (1981). *Guide to the Identification of some of the more Difficult Vascular Plant Species*. Nature Conservancy Council.

WILCOX, M. (2010). 'A novel approach to the determination and identification of *Juncus × diffusus* Hoppe and *J. × kern-reichgeltii* Jansen & Wacht. ex Reichg.' *Watsonia*, **28**: 43–56.

WILLCOX, N. A. & DORE, C. P. (1982). 'A vegetation survey of the Isle of Noss NNR'. Nature Conservancy Council.

WILLIAMSON, K. (1959), *in* Wallace, E. C. (comp.), 'Plant records'. *Proceedings of the Botanical Society of the British Isles*, 3: 181–203.

WILMOTT, A. J. (1942). 'Some remarks on British Rhinanthus'. *Botanical Society and Exchange Club of the British Isles. Report for 1939–1940*, **12**: 361–379.

WYNDE, R. M. (1986). 'The Status of Relict Tree Species in Shetland'. (A report to the Nature Conservancy Council.)

X.Y.Z. (1898). [Two dubious records are cited, one for alpine willowherb (not certainly recorded from Shetland), the other for bulbous buttercup.] *Shetland News*, 3 September 1898: 5.

YEO, P. F. (1970). '*Euphrasia brevipila* and *E. borealis* in the British Isles'. *Watsonia*, **8**: 41–44.

—— (1978). 'A taxonomic revision of *Euphrasia* in Europe'. *Botanical Journal of the Linnean Society*, **77**: 223–334.

YOUNIE, D. & BLACK, J. S. (1979). 'A survey of surface seeded swards in crofting areas of northern Scotland'. Bulletin no. 16. North of Scotland College of Agriculture.

ZEHETMAYR, J. W. L. (1953). 'Experimental plantations in the far north of Scotland'. *Scottish Forestry*, **7**: 71–78.